THE POLITICS OF CRUDE OIL PRICING
IN THE MIDDLE EAST, 1970-1975

RESEARCH SERIES, NO. 31

The Politics of Crude Oil Pricing in the Middle East, 1970-1975

A STUDY IN INTERNATIONAL BARGAINING

Richard Chadbourn Weisberg

INSTITUTE OF INTERNATIONAL STUDIES
University of California
Berkeley

International Standard Book Number 0-87725-131-2
Library of Congress Card Number 77-620041
© 1977 by the Regents of the University of California

CONTENTS

List of Tables	viii
List of Figures	viii
Acknowledgments	ix
Abbreviations	x
INTRODUCTION	1

PART ONE
INTERNATIONAL BARGAINING AND THE PRICE OF MIDDLE EASTERN CRUDE — 3

I. A BARGAINING MODEL OF OIL PRICE DETERMINATION	5
II. THE ECONOMIC PARAMETERS OF THE BARGAINING MODEL	13
A. Patterns in Petroleum Production and Consumption	13
B. Price Movements During 1960-1969	15
1. *The Supply Price*	15
2. *The Market Price*	16
3. *The Reference Price*	19
4. *The Parameters of Price*	19
C. Bargaining and Fiscal Policy	22
D. Conclusions	31

PART TWO
THE MARKET IN DISEQUILIBRIUM — 35

III. THE LIBYAN NEGOTIATIONS OF 1970	37
A. Oil Prices and Libyan Political Objectives	38
B. The Manipulation of Threats	39
C. Strategy and Economic Sanctions	41
1. *Bargaining Strength and Supply Logistics*	43
2. *International Cooperation and Bargaining Strength*	45
3. *Consuming-Country Intervention and Bargaining Strength*	48
D. Capitulation and Its Consequences	50

CONTENTS

IV. THE SHIFTING BALANCE OF MARKET POWER: THE ACCORDS OF TEHERAN AND TRIPOLI — 52
  A. The Structure of the Negotiations — 52
    1. *Company Cooperation and the Structure of Negotiations* — 54
    2. *U. S. Intervention and the Structure of Negotiations* — 56
  B. Final Settlement with the Gulf Producers — 60
  C. The Settlement with the Mediterranean Exporters — 63
  D. Conclusion — 66

V. THE BREAKDOWN OF COMPANY-GOVERNMENT BARGAINING, 1972-1973 — 69
  A. Supply, Demand, and Productive Capacity — 69
  B. Prices and Revenues — 71
  C. The Evolution of Company Objectives — 74
  D. Currency Fluctuations and Tax Increases — 76
  E. Participation Negotiations — 78
  F. Nationalization — 82
    1. *Nationalization in Libya* — 82
    2. *Nationalization in Iraq* — 84
  G. Politics and Production Levels in the Persian Gulf — 86
    1. *Oil and Politics in Saudi Arabia* — 86
    2. *U. S. Policy Response* — 88
  H. Conclusion — 90

PART THREE
THE POLITICS OF PRICE DETERMINATION — 93

VI. THE 1973-1974 ARAB OIL EMBARGO — 95
  A. Postwar Supply Interruptions — 95
  B. The Politics of Embargo — 98
    1. *Imposition of Production Cutbacks* — 99
    2. *Embargo Policy* — 99
    3. *The Evolution of Embargo Policy* — 103
  C. The Role of the Companies During the Embargo — 106
  D. The Price of Crude During the Embargo — 112
  E. Conclusion — 113

CONTENTS

VII. INTERNATIONAL POLITICS AND THE PRICE OF CRUDE, 1973-1975 — 116
   A. Energy Policy in the OECD Area Before October 1973 — 116
   B. Policy Response in the OECD Area — 119
      1. Western Europe and Japan — 119
      2. The United States — 123
   C. Bilateral and Multilateral Responses to Higher Oil Prices — 124
   D. Conclusion — 129

VIII. INTRA-OPEC BARGAINING AND THE PRICE OF MIDDLE EASTERN CRUDE, 1973-1975 — 130
   A. OPEC: Structure and Function — 130
   B. The Political Economy of the OPEC Process — 131
      1. Output Policy in a Producers' Cartel — 132
      2. Output Policies of Individual Producers — 134
      3. Market Structure and the Output Policy of Individual Producers — 137
   C. Bargaining and the Price of Crude — 138
   D. Conclusion — 144

GENERAL CONCLUSIONS — 146

BIBLIOGRAPHY — 149

LIST OF TABLES

1. Energy Consumption, Oil Consumption, and Oil Imports by Source for the United States, Western Europe, and Japan: 1962 and 1972 — 14
2. Basic Terms of Original LPG Offer, Gulf Producers' Demands, and Final Teheran Agreement for 1971-1975 — 61
3. Government Revenue Per-Barrel for Saudi Arabian 34° API Marker Crude: 1970-1973 — 70
4. Real GNP Increases for Developed World: 1970-1973 — 70
5. Sanctions Imposed by OAPEC Members: 6 October-2 November 1973 — 101
6. Non-Communist World Crude Oil Production During Embargo: September 1973-March 1974 — 108
7. Comparative Deliveries of Crude Oil and Products to the Largest Consuming Areas by Five U. S. Major Oil Companies During a Base Period and the Embargo — 111
8. Selected OPEC Countries' Market Shares, Share of Total OPEC Reserves, and Population for 1973; Usable Productive Capacity in 1974 — 136
9. Government Revenue Per-Barrel for Saudi Arabian 34° API Marker Crude: October 1973-January 1975 — 139
10. Crude Oil Production by OPEC Countries: September 1973-December 1974 — 143

LIST OF FIGURES

1. Market Price of the Representative European Crude f.o.b. as Observed from Third-Party Sales: 1960-1970 — 18
2. Payments to Producing Countries and Earnings of Seven Major Oil Companies in the Eastern Hemisphere: 1960-1969 — 20
3. Estimated Non-Communist World Spare Crude Oil Producing Capacity: 1960-1973 — 72

ACKNOWLEDGMENTS

The research for this study was pursued during my tenure from 1974-76 at Nuffield College, Oxford University, as a Marshall Scholar. I would like to thank the Marshall Aid Commemoration Commission and the Warden and Fellows of Nuffield College for their generous support. For their valuable criticism and suggestions in the early stages of my research, I am grateful to Dr. W. M. Corden of Nuffield College and to Mr. P. M. Oppenheimer of Christ Church College, Oxford University. My investigations brought me in contact with a number of executives of the major oil companies, independent oil companies, and investment banks. I am indebted to them for finding the time to give patient and comprehensive answers to my questions.

Part of the research for this study was pursued in Washington, D. C., during the summer of 1975. I was aided considerably by the staff members of the Federal Energy Administration, the Department of the Interior, the Senate Subcommittee on Multinational Corporations, the Library of Congress, and the American Petroleum Institute. I want to acknowledge their kind assistance.

I am grateful to Mr. L. Whitehead and to Dr. K. Hossain, both of Nuffield College, and Mr. A. F. Peters of the Shell International Petroleum Company, London, who read over earlier drafts of this study and made valuable suggestions. Shortly before his death, Professor A. Buchan of Oxford University shared with me some of his vast knowledge of international politics, and contributed important insights. Mr. R. Mabro of the Institute for Economics and Statistics, Oxford University, helped me to improve both the organization and substance of this study. Finally, my thanks are due to Mr. P. Gilchrist of the Institute of International Studies for his careful and effective editorial work.

R. C. W.

April 1977
Yale University
New Haven, Connecticut

ABBREVIATIONS

API	American Petroleum Institute
Aramco	Arabian American Oil Company
b/d	Barrels per day
BP	British Petroleum Company, Ltd.
CFP	Compagnie Francaise des Pétroles
EEC	European Economic Community
ENI	Ente Nazionale Indrocarburi (Italy)
ERAP	Enterprise de Recherches et d'Activités Pétrolièrs (France)
Esso	Standard Oil Company of New Jersey and subsidiaries (now Exxon Company)
Exxon	Exxon Company and subsidiaries (formerly Standard Oil Company of New Jersey)
FEA	Federal Energy Administration
FEO	Federal Energy Office
f.o.b.	Free on board
IEA	International Energy Agency
IPC	Iraq Petroleum Company, Ltd.
LPG	London Policy Group
NIOC	National Iranian Oil Company
OAPEC	Organization of Arab Petroleum Exporting Countries
OECD	Organization for Economic Cooperation and Development
OPEC	Organization of Petroleum Exporting Countries
RCC	Revolutionary Command Council (Libya)
Shell	Royal Dutch/Shell
Socal	Standard Oil Company of California
Tapline	Trans-Arabian Pipeline
UAE	United Arab Emirates

INTRODUCTION

Crude oil exported from Middle Eastern and North African countries dramatically increased in price during the 1970-75 period.[1] The purpose of this study is to attempt to determine the reasons for that development.[2]

The prevailing economic and political models for analyzing commercial relations between states do not provide a satisfactory explanation for the timing or the degree of these price increases. As a result, it has been necessary to develop an alternative model in an attempt to supply such an explanation. This alternative model—the bargaining model—treats price as a variable which reflects the power of the actors in the market. Power is a concept which includes not only the supply and demand factors which comprise the classic model of price determination, but also a number of other elements.

A vast literature on the international petroleum industry exists, but scholarly appraisals of the industry tend to concentrate on the economics of petroleum supply and demand and on the organization of the industry. While a few years ago it would have been difficult to undertake a study of the political elements involved in the determination of oil prices, in recent years the information required to analyze the behavior of the major actors in the petroleum market has become increasingly available. Of the seven major international oil companies, five are domiciled in the United States.[3] These companies have recently been closely scrutinized by the U.S. Congress, and there is now within the public domain sub-

---

[1] Specifically, the oil-exporting countries referred to are Algeria, Iran, Iraq, Kuwait, Libya, Oman, Qatar, Saudi Arabia, and the United Arab Emirates (UAE).

[2] A study of this type is concerned with "empirical" or "causal" theory: it analyzes actual political and economic behavior in order to arrive at a coherent method for understanding a large amount of data. See Stanley Hoffman's discussion of problems of scope, method, and purpose in "Commentary," in *Contemporary Theory in International Relations*, ed. Stanley Hoffman (Englewood Cliffs: Prentice-Hall, 1960), p. 8.

[3] The seven major companies are Exxon, Texaco, Gulf, Socal, Mobil, British Petroleum (BP), and Shell; only BP and Shell are not domiciled in the U.S.

stantial information concerning negotiations between the companies and the governments of the oil-producing countries, as well as the petroleum policies of the U.S. government.[4] Despite the greater availability of information, there are obvious problems in analyzing a recent historical period. My goal in this study has been to use the available evidence to provide an explanation for price developments which is as accurate as is currently possible. Though this investigation is essentially a case study in the pricing of oil in an international market, I hope that it will suggest a useful methodology for analyzing other issues in international economic diplomacy.

---

[4]Raymond Vernon, "The Influence of the U.S. Government upon Multinational Enterprise: The Case of Oil" (address at the Second International Colloquium on the Petroleum Economy, Laval University, Quebec, 3 October 1975). For a summary of the major papers presented at this colloquium, see *Petroleum Economist* (hereinafter cited as *PE*), November 1975, pp. 407-9.

PART ONE

INTERNATIONAL BARGAINING AND THE PRICE OF MIDDLE EASTERN CRUDE

Chapter I

A BARGAINING MODEL OF OIL PRICE DETERMINATION

We shall begin by constructing a political-economic model to explain international cooperation and conflict with regard to the pricing of oil—the single most important commodity in world trade. The prevailing political and economic models of commercial relations between states do not satisfactorily explain recent developments in the market for Middle Eastern crude. The classical *economic* model of international commerce is a theoretical abstraction used by liberal economists in order to give normative guidance to policymakers rather than to explain behavior. The argument for free trade which emerges from the classical economic model is a convincing one, but a number of features of the model undermine its utility as an explanation of trade policy.[1] These features are (a) the acceptance of economic efficiency as the most important objective of trade policy, (b) a reliance upon the methodology of comparative statics, (c) the assumption that markets are competitive, and (d) the implicit assumption of a stable political environment.

Because of these features, the classical economic model of the determination of price of an internationally traded good has little applicability to the world oil market. First, while the classical economic model accepts efficiency as the most important goal of policy, oil is the preeminent strategic commodity. Oil is vital in both war and peace, and policymakers in the oil-consuming countries have traditionally been less concerned with economic efficiency than with assured access to oil supplies. Second, the classical model relies upon comparative statics; however, comparative statics is an unduly restrictive framework for explaining the dramatic increase in oil prices. Third, the classical model assumes that markets are highly competitive, but the oil market has consistently been dominated by a small number of large international firms with the opportunity

---

[1] See Lawrence B. Krause and Joseph S. Nye, "Reflections on the Economics and Politics of International Economic Organizations," in *World Politics and International Economics*, eds. C. Fred Bergsten and Lawrence B. Krause (Washington, D.C.: Brookings Institution, 1975), p. 328.

for collusion. Also, in the absence of competitive markets, distribution in the classical model is indeterminate, but the distribution of oil rents was a basic issue in the conflict over oil prices.[2] Fourth, the classical model implicitly assumes a stable political environment, but the Middle East and North Africa, which were the largest exporters of crude oil during the period under consideration,[3] have been politically unstable. This political instability reflected both political struggles within the Middle Eastern and North African states and the fact that the issues of oil production and price were linked to the Arab-Israeli conflict and United States-Soviet rivalry in the Middle East.

Prevailing *political* models of international relations are also inadequate for analyzing developments in the oil market. Prevailing political models tend to be based upon three assumptions: (a) nation-states are the most significant actors in world politics; (b) military power is the most important instrument of nation-states; (c) security and status are the dominant objectives of nation-states.[4]

These assumptions concerning actors, instruments, and goals appear to be inappropriate for the analysis of international cooperation and conflict with regard to commodity pricing. First, while political models assume that nation-states are the most significant actors, international firms played a key role in the process by which the price of Middle Eastern crude was determined. Second, while political models assume that military power is the most important instrument in the international arena, changes in the nature of military force and in the consequences of the use of such force have seriously limited its utility. Nation-states relied upon other instruments of power and influence in attempts to achieve their oil policy objectives. Third, political models assume that security and status are the dominant objectives of international actors; although security and status have remained essential goals of states, policymakers in the producing states of the Middle East put a higher priority on development, perceiving that increased oil revenues could be the best means of achieving security and status. On the other hand, policymakers in the oil-consuming countries put

[2] The rent on a barrel of oil is the difference between its supply price and the price at which it is sold (see p. 32 below).

[3] Middle Eastern producing countries accounted for 60 percent of all oil moving in international trade in 1974 (*BP Statistical Review of the World Oil Industry 1974* [London: British Petroleum Company, 1974], p. 10).

[4] Krause and Nye, "Reflections on the Economics and Politics," p. 326.

an increasing emphasis upon welfare goals, such as growth, income, and employment.

In any commodity transaction, price depends upon the balance of power on each side of the bargaining table,[5] and the basic assumption of our model is that the price of crude oil was a variable which reflected the power of the various actors in the market. Before we can elaborate our model of price determination, however, we must carefully analyze the concept of power.

Power can be defined as the ability of an actor to achieve objectives when faced by external resistance. It can be divided into economic power and political power.[6] Economic power is based upon a set of economic relationships and political power upon a set of political relationships among actors. Economic power is determined by the characteristics of supply, demand, resource location and market structure specified in the classical economic model of the determination of the price of an internationally traded commodity. In the classical model, economic power is measured by the reciprocal demand that fixes the terms of trade.[7] In our bargaining model, economic power—once obtained—may be used not only to determine price but also to attain noneconomic objectives. Although political power stems from noneconomic relationships, it can be directed toward international economic relations. Four types of political power can be defined: state, military, legal, and collective.[8] State power is based upon the sanctions which a state has at its disposal *qua* state, such as expropriation and nationalization. State power ultimately rests upon military power, which ensures the sovereignty of the state. Military power is based upon the threat or the use of military force. Legal power derives from legal relationships among actors. Within a nation, legal power ultimately rests upon state power, since the state imposes sanctions for the violation of laws. Finally, collective power is derived from

---

[5] J.E. Hartshorn, *Oil Companies and Governments: An Account of the International Industry in Its Political Environment* (London: Faber and Faber, 1962), p. 144.

[6] J. Pen, "Bilateral Monopoly, Bargaining and the Concept of Economic Power," in *Power in Economics: Selected Readings*, ed. K.W. Rothschild (Harmondsworth: Penguin, 1971), pp. 105-6. [Reprinted from J. Pen, *The Wage Rate under Collective Bargaining* (Cambridge, Mass.: Harvard University Press, 1959), pp. 61-105, 207-9.]

[7] Klaus Knorr, "The Limits of Economic and Military Power," *Daedalus* 104 (Fall 1975): 231. [Special issue on "The Oil Crisis: In Perspective," ed. Stephen R. Graubard.]

[8] Pen, "Bilateral Monopoly," pp. 108-9.

an agreement among actors to adhere to a set of rules, and it exists only when the actors adhere to this set of rules.

Within a domestic market, the process of bargaining over prices is defined by the exercise of state power. In a transaction between two American oil brokers, for example, price will be determined by economic factors—in other words, the alternative supplies and markets open to buyers and sellers and their knowledge of such supplies and markets;[9] price in such a transaction will not be determined by the use of military power. On the other hand, in a transaction between Saudi Arabia and the United Kingdom, the price of oil will be determined not only by economic factors, but also by the political and military capabilities and objectives of these countries.

The price of oil in international markets is a variable which reflects both the political and economic power of the actors in the market. In other words, reciprocal demand is determined not only by the characteristics of supply, demand, resource location, and market structure, but also by the exercise of political power. For example, Libya's nationalization of the producing assets of British Petroleum (BP) in 1973 was an act of state power. However, when the Libyan government attempted to market the nationalized oil, it was temporarily denied access to alternative buyers because of BP's legal actions in Italy—in other words, by Italy's exercise of legal power.[10]

A model of price determination must include, at a minimum, the following elements: (a) the principal actors and their objectives, (b) the resources and policies which the actors may employ to achieve their objectives, (c) the process by which conflicting objectives and policies affect the course of events.[11] We shall discuss these elements of our model as follows:

(a) As to principal actors, our model involves three sets—namely, (i) the principal oil-consuming countries (Western Europe, the United States, and Japan), (ii) the international oil companies with producing interests in the Middle East, and (iii) the major oil-exporting countries.

As to objectives, the basic objective of the principal oil-consuming countries was to assure continued access to foreign

---

[9] Hartshorn, *Oil Companies and Governments*, p. 144.

[10] See discussion of nationalization in Libya below, pp. 82-84.

[11] See David Easton, *A Systems Analysis of Political Life* (New York: Wiley, 1965), pp. 25-27.

sources of oil; the basic objective of the international oil companies was to maximize total profit on their integrated operations, and the basic objectives of the oil-exporting countries were to increase the value of their oil resources and to exercise greater national control over their exploitation.

However, there were important differences in objectives among members of the same set of actors. For example, all the major oil-consuming countries sought assured access to foreign supplies of crude, but Western Europe and Japan were more dependent than the United States on Middle Eastern sources, and therefore accorded this objective a higher priority than did the United States. All the international oil companies sought to increase the return on their assets; however, there were important differences between the objectives of the independent companies without diversified sources of supply and the objectives of the major international companies (majors). The majors sought to increase their margins by restricting supply to demand at the prevailing price structure; in contrast, the independent companies sought to increase their total profits by expanding their market shares. Finally, all the oil-exporting countries sought to increase the value of their resources, but the *Arab* exporting countries also sought to use oil production as a weapon to influence U.S. policy in the Middle East.

(b) As to resources and policies, there were important differences among the three sets of actors. For example, the military capabilities of the oil-consuming countries were far greater than those of the oil-exporting countries. There were also important differences within each set of actors. The capabilities of the majors, for example, to draw on alternative sources of supply or to sell in alternative markets were far greater than the similar capabilities of the independent companies.[12] We submit that the policies pursued by states and by oil companies to increase their collective power were the most important political determinants of price.[13] More specifically, the collective efforts of the oil companies and of the governments of oil-producing countries to manage crude oil production were the controlling determinant of price.

[12] See George Lenczowski, "Multinational Oil Companies: A Factor in Middle East International Relations," *California Management Review* 8 (Winter 1970): 44.

[13] Although the behavior of companies in a concentrated industry is usually the subject of industrial economics, we have noted that collective power is based upon a set of rules. The process by which such rules are made and enforced is in essence a political process and can be analyzed usefully from the perspective of political science.

(c) As to the process by which price was determined, our model assumes explicit bargaining among the actors.[14] It is assumed that the ability of an actor to attain objectives depends to an important degree upon the decisions and policies of the other actors.[15] It is also assumed that bargaining involves issues in which the actors have opposing interests as well as issues in which the actors have a common interest.[16] All three sets of actors shared an interest in continued oil production and trade. Conflict among the three sets of actors centered upon (i) the level of producing government taxes, which largely determined the amount and distribution of oil rent, and (ii) control over the production and transportation of crude oil. In our model, conflicting objectives and policies could be reconciled by two mechanisms. First, the actors could attempt to change the perceptions of the opponents by threats or promises; second, the actors could attempt to coerce their opponents by implementing various sanctions.

Our final assumption is that the actors enter into explicit bargaining in response to economic changes that influence price. According to our model, if the price of crude increases in response to an increase in demand, the companies, producing governments, and consuming countries will negotiate on price. The model predicts that (i) producing governments will seek to obtain higher revenues, (ii) companies will resist higher producing government revenues *if* they believe that higher revenues will lower profits and (iii) consuming governments will resist higher producing government revenues *if* they believe that such revenues will diminish their assurance of access to supplies of crude oil.

This model of oil pricing differs significantly from the models of most industrial economists, who attempt to explain the level of price by formulating direct empirical links between the structure of a given market and the performance of companies.[17] For example, such a model would predict that given the rapid shift of total non-Communist world oil production to the Middle East, the evolution of the major exporting countries into a cartel was inevitable. However, such a model does not explain how the production policies of the exporting countries were coordinated. This

[14]Thomas C. Schelling, *The Strategy of Conflict* (Cambridge, Mass: Harvard University Press, 1963), p. 21.

[15]*Ibid.*, pp. 15-16.

[16]*Ibid.*, p. 14.

[17]In particular, see Joe S. Bain, *Industrial Organization* (New York: Wiley, 1950). Bain claims that the inclusion of conduct variables is not essential to the development of a general theory of industrial organization.

is a question to which economic theory provides no answer. The so-called theory of imperfect competition is only an explanation of how market power will be exercised once it is obtained.[18] In our view, market structure does not provide an adequate explanation of the behavior of actors in the market, since any given market structure is compatible with a wide variety of price levels.[19]

Our model of interdependent bargaining, as developed so far, provides only an explanatory framework in which cooperation and conflict in oil pricing can be analyzed. In the remainder of this study, we shall analyze the role of politics and economics in the bargaining over price which took place during the periods 1960-69, 1970-73, and 1973-75. These three periods differ from each other in three respects: (a) the distribution of power among actors, (b) the involvement of actors in the negotiations, and (c) the use of sanctions in bargaining.

The period 1960-69 was characterized by bilateral negotiations between companies and producing governments over per-barrel excise taxes. Negotiations over price were generally treated as nonpolitical commercial disputes, and were settled without recourse to sanctions. The collective power of the companies to deal with producing governments that made excessive demands for revenue, by shifting production to other producing states, balanced the collective power of the producing states to increase their revenue by concertedly reducing production. However, the ability of the major oil companies to act in concert declined during 1960-69 because of (a) the expansion of the operations of the smaller independent companies and (b) the rapid increase of the proportion of Middle Eastern and North African output in total non-Communist world consumption.

The beginning of the period 1970-73 was marked by the imposition of production cutbacks by the Libyan government. The use of this sanction, and the inability of the companies to make up production elsewhere, signalled a fundamental shift in the balance of power in favor of the producing countries.[20] This period was

---

[18]Harold Demetz, *The Market Concentration Doctrine: An Examination of Evidence and a Discussion of Policy* (Washington: American Enterprise Institute, 1973), p. 26.

[19]M.A. Adelman, " 'World Oil' and the Theory of Industrial Organization," in *Industrial Organization and Economic Development*, eds. Jesse W. Markham and Gustav F. Papanek (Boston: Houghton Mifflin, 1970), p. 151.

[20]J.E. Hartshorn, "From Tripoli to Teheran and Back: The Size and Meaning of the Oil Game," *The World Today* 27 (July 1971): 291.

characterized by the greater involvement of consuming countries in price negotiations, and by the linking of oil policy to wider political issues.

The beginning of the period 1973-75 was marked by the use of the Arab "oil weapon." The use of production cutbacks as an instrument of foreign policy resulted in the evolution of the Organization of Petroleum Exporting Countries (OPEC) into an effective governmental cartel.[21] Despite direct negotiations between consuming and producing governments, the price of crude was determined by bargaining among OPEC members.

---

[21]The economic model usually evoked by the term "cartel" involves the setting of market shares. (See R. Mabro, "Can OPEC Hold the Line?," Supplement to *Middle East Economic Survey*, 28 February 1975, pp. 1-2.) From 1973-75, OPEC members agreed only to set price and not output shares. When we refer to OPEC as a cartel, we mean a cartel of nation-states which maintains market discipline by a formula pricing scheme. This is a variant of what Scherer calls "Rule-of-Thumb pricing" (F.M. Scherer, *Industrial Market Structure and Economic Performance* [Chicago: Rand McNally, 1970], pp. 173-79).

Chapter II

THE ECONOMIC PARAMETERS OF THE BARGAINING MODEL

In our bargaining model of oil price determination, economic factors (a) partially explain the interests of actors in price negotiations, (b) provide one set of determinants of the outcome of such negotiations, and (c) set the lower and upper limits to oil prices. This chapter deals with oil supply, demand, and price, and is divided into three sections. In the first section, shifts in the magnitude and distribution of oil production and consumption are examined; in the second section, some of the difficulties in determining an "economic" price for Middle Eastern crudes—difficulties that stem from the structure of the oil market—are considered; in the third section, price movements in the 1960-69 period are analyzed in terms of bargaining between the companies and the producing governments.

A. PATTERNS IN PETROLEUM PRODUCTION AND CONSUMPTION

Energy consumption in the non-Communist world during the post-World War II period has had three particularly important characteristics. First, there has been a relatively rapid increase in the total consumption of energy. (The rate of growth in energy consumption during the period 1960-72 was approximately 5.5 percent per annum.)[1] Second, there has been a switch to oil, largely at the expense of coal. (In 1960 oil accounted for 35.8 percent of total eneregy consumption and coal accounted for 44.2 percent; in 1972, however, oil accounted for 46.0 percent of the total and coal for only 28.7 percent.)[2] Third, there has been an increasing proportion of Middle Eastern exports in total oil consumption. Table 1 shows the development of the Middle East into the dominant source of oil

---

[1] U.S. Federal Energy Office, Office of International Energy Affairs, *A Discussion of the World Energy Market in 1980 and 1985* (Washington: Federal Energy Office, April 1974), p. 9, table 5.

[2] Joel Darmstadter and Hans H. Landsberg, "The Economic Background," *Daedalus* 104 (Fall 1975): 19, table 2.

THE POLITICS OF CRUDE OIL PRICING

Table 1

ENERGY CONSUMPTION, OIL CONSUMPTION, AND OIL IMPORTS BY SOURCE FOR THE UNITED STATES, WESTERN EUROPE, AND JAPAN: 1962 AND 1972

	1962			1972		
	U. S.	Europe	Japan	U. S.	Europe	Japan
	Million Barrels per Day					
Energy consumption (b/d oil equivalent)	23.27	13.96	2.25	35.05	23.84	6.58
Oil consumption	10.23	5.24	.96	15.98	14.20	4.80
Oil imports[a]	2.12	5.19	.98	4.74	14.06	4.78
From Middle East/ North Africa	.34	3.80	.72	.70	11.30	3.78
Other areas	1.78	1.39	.26	4.04	2.76	1.00
	Percentage of Energy Consumption					
Oil consumption	44.0	37.5	42.7	45.6	59.6	73.0
Oil imports[a]	9.1	37.2	43.6	13.5	59.0	72.6
From Middle East/ North Africa	1.5	27.2	32.0	2.0	47.4	57.4
Other areas	7.6	10.0	11.6	11.5	11.6	15.2
	Percentage of Oil Consumption					
Oil imports[a]	20.7	99.0	102.1	29.7	99.0	99.6
From Middle East/ North Africa	3.3	72.5	75.0	4.4	79.5	78.6
Other areas	17.4	26.5	27.1	25.3	19.4	20.9

Source: Darmstadter and Landsberg, "Economic Background," p. 21, table 4.

[a] Oil imports are gross, and exclude product exports and refinery losses. Thus, the degree of foreign dependence is slightly overstated for Europe, and the figures show an excess of imports in relation to consumption for Japan.

ECONOMIC PARAMETERS OF THE BARGAINING MODEL

supplies for Western Europe and Japan, as well as the rising percentage of Middle Eastern exports in U.S. consumption. By 1972 the Middle East had reached a clearly dominant position with respect to Western European and Japanese oil supplies, accounting for 57.4 percent of total energy consumption in Japan and 47.4 percent of total energy consumption in Western Europe. The amount of crude oil moving in world trade increased from 6.4 million b/d in 1969 to 24.9 million b/d in 1972,[3] reflecting an increasing shift to production in areas geographically removed from consumption.

B. PRICE MOVEMENTS DURING 1960-1969

Price in economic theory is not one but three distinct concepts: (1) supply price—the least that needs to be paid to elicit a given stream of output, (2) market price—the current exchange rate of money against a good, and (3) reference price—the price which measures the expected benefit of an investment against its cost.[4] Since investment is made to produce a profit, a decision to invest means that the present value of expected revenues will exceed or at least equal the supply price.[5]

1. The Supply Price. There are a number of studies on the supply price of oil—the price that has to be paid to nature in order to produce a barrel of crude. The results vary because the studies are based upon different assumptions and methods. Bradley estimates that average development costs per barrel of output during 1953-62 were 11-15 cents[6] for the major Middle Eastern producing areas, assuming a 15 percent annual rate of return; his estimate for Libya is 21.9 cents/bbl. and for Venezuela 39.4 cents/bbl.[7] Adelman estimates development costs per-barrel at 4.7 cents for Iran during

---

[3] Neil H. Jacoby, *Multinational Oil: A Study in Industrial Dynamics* (New York: Macmillan, 1974), p. 59, table 4.4.

[4] M.A. Adelman, *The World Petroleum Market* (Baltimore: Johns Hopkins, 1972), p. 195.

[5] In other words: "Expenditures for capacity will not be made if the incremental producing cost exceeds the price of crude" (Paul G. Bradley, *The Economics of Crude Petroleum Production* [Amsterdam: North Holland, 1967], p. 105).

[6] In accordance with industry convention, the American system of weights and measures is used in this study; likewise, prices and costs are stated in U.S. dollars.

[7] Bradley, *Crude Petroleum Production*, p. 102, table 6.6.

1963-69 and 4.1 cents for Saudi Arabia during 1966-68; these compare with his estimates of 7.4 cents for Libya during 1966-68, 35.1 cents for Venezuela during 1966-68, and $1.04 for the United States during 1960-63. All of Adelman's estimates are based upon the assumption of 20 percent annual rate of return on investment.[8]

The Adelman and Bradley studies, as well as other economic studies of the marginal cost curve in the Middle East, agree that development costs would not rise appreciably if production were greatly expanded.[9] In 1970 the absolute cost per-barrel for crude oil in the Middle East was so low that even large percentage errors would not make a great difference. Development costs and operating costs were both under 5 cents; thus the total cost of an incremental barrel of Middle Eastern crude oil in the cheapest producing area was about 10 cents.[10] During 1960-69 these costs were about one-tenth of the market price.[11]

2. The Market Price. The market price represents the exchange rate between money and a barrel of crude. However, since most crude oil in the world market remained within the integrated channels of the international companies, most of the data on oil prices had nothing to do with market price at all. In an integrated company, the price at which crude oil was transferred from a producing affiliate to a refining affiliate determined the distribution of total company profit between crude production and refining.[12] Thus the higher the transfer price of crude, the lower the (pre-tax) profit

---

[8] Adelman, *World Petroleum Market*, p. 76, table II-8.

[9] Charles Issawi, *Oil, the Middle East, and the World* (Beverly Hills: Sage Publications, 1972), p. 24.

[10] M.A. Adelman, "Oil Demand, Supply, Cost and Price in the World Market," in United Nations, Department of Economic and Social Affairs, Ad Hoc Panel of Experts on Projections of Demand and Supply of Crude Petroleum and Products, *Petroleum in the 1970s* (ST/ECA/179), 18 March 1971, pp. 166-67. These estimates are widely accepted and have been independently substantiated. The U.S. Tariff Commission estimates that real extraction costs—that is, development and operating costs per-barrel—in the Persian Gulf were 12 cents in January 1971 (U.S. Congress, Senate, Committee on Finance, *World Oil Developments and U.S. Oil Import Policies*, 93rd Cong., 1st sess., p. 26, table 7).

[11] P.H. Frankel, "Fossil Fuel Developments—The Financial Implications," address at the Conference on World Energy Finance, London, 7 October 1974.

[12] Edith T. Penrose, *The Large International Firm in Developing Countries: The International Petroleum Industry* (London: George Allen and Unwin, 1968), p. 186.

attributed to refining and distribution, with total profits on integrated operations remaining the same. If there were any differences in the rates of taxation on various operations, the "accounting cost" of crude would be adjusted in order to minimize the total tax burden of the company, and the transfer price of crude within integrated channels would bear no relation to the market price.

A review of the entire literature on tax policies of the U.S. and British governments as they affected the foreign operations of international oil companies is not within the scope of this study. In the simplest terms, a number of provisions in U.S. tax laws (allowing deductions from income tax for the depletion of oil deposits, deductions for income tax payments to foreign governments, and deductions for the expense of intangible drilling costs) created important tax advantages in attributing profits to the production of crude within the integrated oil companies.[13]

There are two methods of obtaining market price data for Middle Eastern crude oil. First, we can observe prices for crude oil sold by one nonaffiliated company to another—so called "arm's-length" or "third-party" transactions. Second we can calculate the netback value of crude oil f.o.b.[14] Middle East by taking refined product prices in Western Europe in arm's-length sales, less the cost of refining, less the appropriate tanker rate. Adelman's work uses both methods with the same result.[15] Arm's-length data are used in this study. The basic trend in the f.o.b. price of Middle Eastern crude during 1960-70 is shown in Figure 1, substantiating the observation that Middle Eastern crudes sold for not less than ten times their costs[16] during 1960-69. What accounts for this difference between supply cost and market prices? In 1970, about 70 percent

---

[13] A good non-technical summary of the U.S. tax law and its effects is U.S. Library of Congress, Congressional Research Service, Economics Division, *Special Provisions of the Federal Income Tax Affecting the Oil and Gas Industry*, by Jane Gravelle, Publication 72-189E (Washington: Library of Congress, 25 August 1972). Both the percentage depletion allowance and intangible drilling cost allowance were restricted by 1974 legislation. In 1974, the Federal Energy Administration was also given additional authority to monitor the prices of crude oil that was transferred within integrated corporate channels to U.S. affiliates, in order to prevent transfer pricing and tax avoidance.

[14] "f.o.b." means free on board—that is, the price quoted to load a barrel of oil on a tankship, after which the buyer becomes responsible for all freight charges.

[15] Adelman, *World Petroleum Market*, p. 183, table VI-3.

[16] The term "cost" as used here includes an adequate return on invested capital.

Figure 1

MARKET PRICE OF REPRESENTATIVE EUROPEAN CRUDE (31° API) f.o.b. AS OBSERVED FROM THIRD-PARTY SALES: 1960-1970

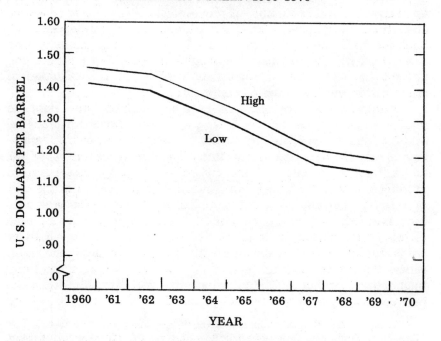

Note: Adelman calculated the representative crude in Western Europe as 31° API on the basis of actual refinery runs (Adelman, *World Petroleum Market*, p. 183).

Source: Adapted from Standard Oil Company of California Submission for the Record in Extension of Oral Testimony Taken on the Afternoon of March 28, 1974, before the Senate Subcommittee on Multinational Corporations, in U. S. Congress, Senate, Committee on Foreign Relations, Subcommittee on Multinational Corporations, *Multinational Corporations and U. S. Foreign Policy*, Hearings, 93d Cong., 1st and 2nd sess. (Washington: GPO, 1975) 8:655, Exhibit 1. High and low bids for 31° API Arabian Medium estimated by Chevron Oil Trading Company from its sales and public bids.

ECONOMIC PARAMETERS OF THE BARGAINING MODEL

of the difference between the supply cost and the market price was reflected in the revenues obtained by the producing countries.[17] Such revenues were calculated by means of an accounting price for crude—the so-called "posted price." The posted price was subject to various discounts; the sum of the posted price, less discounts, was subject to a tax on putative profits, and the profits tax plus the royalties yielded the host government revenues. Despite the complexity of the system, the foregoing taxes amounted to a per-barrel excise tax. The posted price thus bore no relationship to the market price; it was merely an accounting convention.

Figure 2 shows that producing governments' per-barrel revenues increased during 1960-69; it also shows that companies' profits generated by each barrel of oil sold *decreased*. Thus producing governments were able to raise their per-barrel taxes even at a time when oil prices were declining; however, the decline in market prices could not continue unchecked because the per-barrel tax established a floor price.[18]

3. The Reference Price. The reference price measures the expected benefit of investment against its cost. The oil company executive backs his assessment of the future market price and production cost of oil with scarce capital. Expenditures for oil exploration and production facilities in the Middle East remained relatively constant throughout the 1960-69 period. It continued to be profitable for companies to search for and produce Middle Eastern crudes because expected market prices, less expected tax-paid costs, left a margin large enough to justify a steady stream of new investment.[19]

4. The Parameters of Price. At this point, we can sum up the economic parameters of bargaining on crude oil prices. The price of any commodity has a floor, which under competitive conditions

---

[17] Since the f.o.b. market price for the representative crude was about $1.20/bbl., and taxes about 85 cents/bbl., taxes represented around 70 percent of the market price.

[18] George Polanyi, "The Taxation of Profits from Middle East Oil Production: Some Implications for Oil Prices and Taxation Policy," *Economic Journal* 76 (December 1966): 773.

[19] Capital expenditures on production averaged 280 million dollars a year in the Middle East during 1960-70 (Chase Manhattan Bank, Energy Economics Division, *Capital Expenditures of the World Petroleum Industry 1970* [New York: Chase Manhattan Bank, 1971], pp. 24-25).

Figure 2
PAYMENTS TO PRODUCING COUNTRIES AND EARNINGS OF SEVEN MAJOR OIL COMPANIES IN THE EASTERN HEMISPHERE: 1960-1969

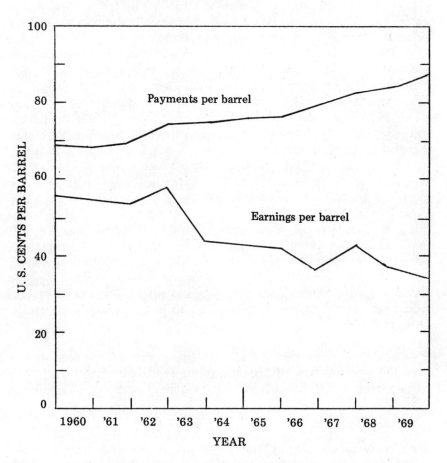

Sources: Adapted from First National City Bank, Petroleum Department, *Energy Memo*, January 1975, pp. 1-4, and Geoffrey Chandler, *Oil—Prices and Profits* (London: Foundation for Business Responsibilities, 1975), p. 11.

ECONOMIC PARAMETERS OF THE BARGAINING MODEL

is the supply cost or the long-term cost of making it available. In 1970 the long-term cost of making additional crude oil available in the Middle East was about 10-15 cents.

The price of any commodity also has a ceiling. In the long run this is set by the price of an alternative source of satisfying demand.[20] If we assume the existence of a monopolistic producer of oil in the Middle East, that producer would set price at a level where the revenue from an additional barrel of oil sold would equal the direct costs of production and the present value of income foregone in the future, since a barrel of petroleum sold today cannot be sold tomorrow.[21] Thus a reasonable estimate of the economic floor and ceiling to the price of Middle Eastern crude might be in the range of 15 cents to 15 dollars per-barrel f.o.b. Persian Gulf. This estimate reflects the extreme assumptions of perfect competition, on the one hand, and complete control of the market, on the other. The actual performance of producing governments during 1960-69 not only indicated some coordination of fiscal policy but also the potential for a great deal more. By the end of the decade, the taxes per-barrel fixed by producing governments were the key factors in setting market price. Such taxes could only be fixed by a cooperative international effort.[22] If any country exacted per-barrel payments much above the average, it would soon find its oil production curtailed and other governments receiving additional revenues from their increased production. By 1970, the concentrated structure of the international petroleum industry was no longer an important factor in explaining the level of price. If the market were perfectly competitive, prices would have decreased by perhaps 20 cents per-barrel, which represented the monopoly rent of the companies.[23]

[20] This is the backstop technology, which—depending on the authority—is liquified coal, shale oil, natural gas liquids, or the U^{238} breeder reactor. Estimates are in the 8-20 dollar per-barrel of oil equivalent range.

[21] See Douglas R. Bohi and Milton Russell, *U.S. Energy Policy: Alternatives for Security* (Baltimore: Johns Hopkins, 1975), p. 39.

[22] This coordination of fiscal policy did not necessarily have to be active, but could have been tacit. For example, all that would be necessary would be a decision by all producing governments to grant new concessions only if terms were comparable to those of already existing concessions in other countries.

[23] Adelman, " 'World Oil,' " pp. 148-49. Nordhaus's quantitative model of oil prices gives results which are in general agreement with Adelman's figure. His calculation of the companies' monopoly rent is 25 cents/bbl. in 1970 (William D. Nordhaus, "The Allocation of Energy Resources," *Brookings Papers on Economic Activity*, No. 3 [1973], p. 559).

C. BARGAINING AND FISCAL POLICY

In August 1960, the majors reduced posted prices unilaterally by 8-10 cents/bbl., which resulted in a loss of some $300 million in tax revenues to Middle Eastern governments.[24] Accordingly, the governments of Iran, Iraq, Kuwait, Saudi Arabia, and Venezuela founded an organization (OPEC)[25] to stabilize crude oil prices and maintain them "steady and free from all unnecessary fluctuations."[26] OPEC's founders intended the organization to be a governmental cartel,[27] and its first resolution directs members to formulate a system of controlling production.[28] OPEC had little success in establishing a joint production program during 1960-69 even though the organization spent over two years (June 1965-June 1967) on the problem. However, the organization did have an important effect on the level of taxes imposed by producing governments. Also, the OPEC technique of collective negotiations on taxes and royalties had long-term significance.

During 1960-69 the producing governments managed to cut progressively the discounts from posted prices, thereby increasing both the putative profit subject to tax and per-barrel revenues. In the fourth OPEC conference held in Geneva in 1962, the organization established three goals: (a) restoration of posted prices, (b) expensing of royalties, and (c) elimination of market allowances.[29]

The market allowance issue was settled by negotiations between the Aramco consortium and Saudi Arabia. The settlement reduced marketing deductions from 4.2 to 0.5 cents/bbl. The establishment

---

[24] Zuhayr Mikdashi, *The Community of Oil-Exporting Countries: A Study in Governmental Cooperation* (London: George Allen and Unwin, 1972), p. 33.

[25] OPEC members now also include Algeria, Ecuador, Indonesia, Libya, Nigeria, Qatar, and the UAE. Gabon is an associate member.

[26] *Resolutions Adopted at the Conference of the Organization of Petroleum Exporting Countries*, Resolution I.1. (Roman numerals indicate the number of the conference and Arabic numerals the number of the resolution.)

[27] Zuhayr Mikdashi, "Cooperation Among Oil Exporting Countries with Special Reference to Arab Countries: A Political Economy Analysis," *International Organization* 28 (Winter 1974): 6.

[28] "Members shall study and formulate a system to ensure the stabilization of prices by, among other means, the regulation of production" (*OPEC Resolutions* I.1).

[29] *OPEC Resolutions* IV.32, 33, 34.

of OPEC increased the bargaining power of the producers; it accelerated intergovernmental coordination, and heightened the determination of governments to realize for themselves any tax advantages granted to any other producing country.[30]

Consultations among producing governments ended the ability of companies to avoid taxes through the "exploitation of ignorance."[31] The consortium in Iran soon reached a settlement on market allowances similar to that of Saudi Arabia. The cost to the companies of complying with producing government demands, however, was relatively slight. Taxes on profits paid to foreign governments could be taken as a credit against U.S. taxes, in accordance with the principle of avoiding double taxation.[32] Additional taxes on "profits" imposed by Middle Eastern governments, therefore, could be offset against U.S. taxes.[33]

The second issue—the expensing of royalties—was resolved by protracted negotiations during 1962-65. Originally, royalties (and in some cases rents for concession areas) were deducted from the amount of the tax on putative profits, and only the balance was

[30] See Mikdashi's discussion of bargaining over terms of oil concessions in Zuhayr Mikdashi, *A Financial Analysis of Middle Eastern Oil Concessions: 1901-1965* (New York: Praeger, 1966), p. 265. OPEC negotiators became more aware of the complexities of accounting procedures and their implications for government revenue after the establishment of OPEC (George W. Stocking, *Middle East Oil: A Study in Political and Economic Controversy* [London: Allen Lane, 1970], p. 364).

[31] The term is used by Geoffrey Chandler, "The Myth of Oil Power: International Groups and National Sovereignty," *International Affairs* 46 (October 1970): 714. Chandler argues that companies should not attempt to avoid producing government taxes through the exploitation of ignorance, because the oil business is long-term in nature.

[32] The 50-50 sharing of net profits instituted in Saudi Arabia in 1951 was the first example of this scheme in the Middle East. It rapidly became the standard. This method of taxing the companies was admittedly designed to divert income tax payments from the U.S. government to the producing governments of the Middle East (Mikdashi, *Financial Analysis*, pp. 148-51). In every year since 1962 the aggregate value of foreign tax credits available to the American petroleum industry has been greater than U.S. taxes on their foreign income (Glenn F. Jenkins, "Tax Preferences and the Foreign Operations of the U.S. Petroleum Industry," in U.S. Congress, House, Committee on Ways and Means, *"Windfall" or Excess Profits Tax*, Hearings, 93rd Cong., 2d sess., 1974, pp. 534-55).

[33] Although this affected only U.S. companies, similar advantages were available under the tax laws of other consuming countries (Sam H. Schurr and Paul T. Homan, *Middle Eastern Oil and the Western World: Prospects and Problems* [New York: American Elsevier, 1971], p. 120).

paid. OPEC members sought to obtain treatment of the royalty, generally at 12.5 percent, as a deduction from posted prices before arriving at the putative profit on which producing governments imposed a 50 percent tax. OPEC governments thus wanted companies to pay the royalty *in addition* to the tax on profits. The potential costs—particularly to the American companies—if such new arrangements were adopted would be substantial. Under U.S. tax law, a foreign tax on profits could be credited against U.S. taxes; a royalty, however, was only deductible as a business expense.[34]

The offers and counter-offers made by the companies and governments indicate a great deal about the bargaining process. Throughout the negotiations, the major companies refused to accept the principle of collective negotiations with OPEC.[35] A consortium of major companies that faced an individual producing country could threaten to shift production to other countries where the tax-paid cost per-barrel was lower. Since the credibility of such a threat depended in part upon prevailing market conditions, the surplus of crude at existing market prices during most of the 1960-69 period gave companies an important source of leverage. On the other hand, by negotiating collectively, producing governments could increase *their* leverage by agreeing not to permit shifts in production; tax changes agreed upon would affect tax-paid costs of crude more or less uniformly in the various producing countries.

In the royalty-expensing negotiations, OPEC contended that oil was an exhaustible asset, and that compensation for its value should be over and above a tax on the profitability of company operations. Moreover, they claimed that deducting royalty payments from tax on putative profits negated the principle of 50-50 profit-sharing between companies and governments.[36] The companies belonging to the Iranian consortium offered in late 1963 to accept the principle of expensing royalties but not to increase per-barrel taxes. The offer was clearly tactical, but OPEC's rejection of the offer shows that, despite the rhetoric of OPEC negotiators, the real issue involved was government revenue. The companies also wanted assurances from OPEC members that (a) the taxes established

[34] See Gerald M. Brannon, *Energy Taxes and Subsidies* (Cambridge, Mass.: Ballinger, 1974), pp. 91-99, and Library of Congress, *Special Provisions of the Federal Income Tax*, p. 12.

[35] Stocking, *Middle East Oil*, pp. 350-60.

[36] Muhamad A. Mughraby, *Permanent Sovereignty Over Oil Resources: A Study of Middle East Oil Concessions and Legal Change* (Beirut: Middle East Research and Publishing Center, 1966), pp. 141-42.

for the Iranian consortium would serve as an industry-wide ceiling and (b) OPEC demands to raise posted prices would be abandoned.

At this stage of the negotiations—in late 1963—the producing countries could have escalated their threats and even resorted to sanctions. It seemed unlikely, however, that the producing countries would be able to cooperate on controlling production. Most countries saw higher revenue as a function of higher production.[37] In the event of a direct government-company confrontation, the companies could have divided and ruled. This confrontation never occurred. The position of the majors on the royalty-expensing issue changed in response to developments in Libya, which we shall now discuss.

The 1955 Petroleum Law of Libya had been jointly drafted by government advisors and oil company representatives. The independent companies were particularly anxious to obtain producing interests in North Africa, and these companies played an important role in drafting the law.[38] The tax-paid cost of crude in the Middle East and North Africa was well below the market price, and obtaining its own producing interests offered very attractive prospects to any company. The independents in Libya soon found oil, and sold it at slightly less than going market prices in order to increase their market shares. The majors, on the other hand, were always reluctant to accept lower-than-current prices, since this would tend to weaken prices everywhere and would cost an important fraction of current revenues. Although the independents were responsible only in part for expanded supply and downward pressure on marketing prices in the 1960-69 period, the majors believed that the independents threatened to disrupt price discipline in the entire industry.[39]

The tax system in Libya made it easy for the independents to cut prices. For tax purposes it was possible for petroleum concessionaires to deduct "marketing expenses," which were liberally defined as all deductions that an operating company thought necessary to sell its oil. Thus the Libyan government, as

[37] During 1960-69, OPEC members consistently adopted policies designed to promote exports, despite ongoing discussions on production controls (Mikdashi, *Community of Oil-Exporting Countries*, p. 47).

[38] Ruth First, *Libya: The Elusive Revolution* (Harmondsworth: Penguin, 1974), p. 189.

[39] For example, a Shell chief executive claimed in 1965 that as a result of the disruptive activities of the independent companies, quantities of "uncommitted oil overhung the market" (*Petroleum Intelligence Weekly* [hereinafter cited as *PIW*], 31 May 1965, p. 5). Another important factor putting downward pressure on market prices was oil sales by the Soviet Union and the Communist bloc countries.

well as the independent companies, was willing to sacrifice per-barrel revenues in order to make a larger return on greater volume.[40]

The offers of the majors to expense royalties reflected a strategy of preventing price shading by the independents with access to Libyan crude. More specifically, the majors offered to pay higher per-barrel revenues if producing governments would harmonize tax-paid costs, and thereby prevent the independents from using Libyan tax provisions to expand supply and gain a larger share of the market.[41] In November 1963, the majors not only accepted the principle of expensing royalties, but for the first time agreed to accept a slight increase in tax burden. The companies offered to expense royalties at 12.5 percent in return for an agreement to allow a discount of 8.5 percent from posted prices. (Libya had joined OPEC in 1962, and the majors wanted to apply this formula to Libya.)

OPEC negotiators rejected the November 1963 proposal of the majors because the increment in per-barrel taxes would have been slight. Although the majors accepted the principle of royalty-expensing, there was no agreement on the increase in per-barrel taxes, and negotiations continued throughout 1964. At the sixth OPEC conference, held in Geneva in July 1964, OPEC defined for its members the minimum requirements of an acceptable offer by the companies. The OPEC conference also decided that, subject to certain modifications, the major companies' most recent proposal would meet these demands. This was made by the companies in July 1964 and involved the expensing of royalties on the basis of an 8.5 percent discount from posted prices, which would decrease in successive years. It would result in an increase in government revenues of 4.5 to 5 cents/bbl., although OPEC demands were reported at 10 cents/bbl.

At OPEC's seventh conference, in November 1964, the organization proved incapable of reconciling the conflicting interests of its members. Iran, Kuwait, Libya, Qatar, and Saudi Arabia expressed their intention of accepting the substance of the majors' latest

[40] In 1964, Esso, the leading major in Libya, used the posted price of $2.22/bbl. as an internal transfer price, thus paying the government about 90 cents/bbl. At the same time, the Oasis group, consisting largely of independents, used an internal transfer price of $1.55/bbl., thus paying the government about 30 cents/bbl. (Stocking, *Middle East Oil*, p. 375). The majors, notably Esso and Mobil, heavily discounted posted prices, but only in connection with the small proportion of their output sold to third parties (Penrose, *The Large International Firm*, p. 205).

[41] See Stocking, *Middle East Oil*, p. 379.

proposal. Iran, Kuwait, and Saudi Arabia—the largest Persian Gulf[42] producers at the time—apparently believed that 5 cents/bbl. was a substantial gain in light of the existing surplus of crude at prevailing market prices. The Libyan government supported the agreement because it realized that the quality and location of its crude made its earlier concession terms uniquely generous, and it wanted a share of the high profits of the operating companies—or, as Adelman has put it, "A rich discovery means a dissatisfied landlord."[43] (The Libyan discovery was very rich. Libyan oil yields a high proportion of high value products such as gasoline and heating oils, and it has a low sulphur content. Also, Libya's location gives it a substantial freight advantage over Persian Gulf crudes.)[44]

The opposition in the seventh OPEC conference to the offer of the companies to expense royalties gradually was not based upon its financial terms: it was based upon issues of principle. Most of the opposition came from Iraq. Iraq rejected the companies' offer on the grounds that its nonfinancial provisions infringed upon Iraqi sovereignty. Since the enactment of its petroleum law of 1961, Iraq had been subjected to various pressures from the operating companies.[45] The companies had demanded that Iraq negotiate its limitation of exploitation of oil resources to an area which was about 0.5 percent of the original concession granted to the Iraq Petroleum Company (IPC). After the revolution of 1958, Iraq had signed a technical aid agreement with the Soviet Union, had withdrawn from the sterling area, and had moved outside the political sphere of the other OPEC members.[46] Along with Iraq, Venezuela and Indonesia objected to provisions of the companies' offer requiring producing governments to abandon demands for raising posted prices and to pledge not to restrict the production or movement of petroleum. (It might be noted that Venezuela and Indonesia had royalty-expensing agreements, and could afford to be mindful of principle.)

[42] The term "Persian Gulf" or "Arabian Gulf," as used herein, includes the following countries: Iran, Kuwait, Saudi Arabia, Oman, Qatar, and the UAE.

[43] Adelman, *World Petroleum Market*, p. 42.

[44] Michael Field, "Oil in the Middle East and North Africa," in *The Middle East and North Africa 1975-1976* (22d ed.; London: Europa Publications, 1975), p. 82.

[45] Abbas Alnasrawi, "Collective Bargaining Power in OPEC," *Journal of World Trade Law* 7 (March-April 1973): 191.

[46] Joe Stork, *Middle East Oil and the Energy Crisis* (New York: Monthly Review Press, 1975), pp. 102-6.

The seventh OPEC conference in November 1964 was unable to reconcile the objectives of the government in Iraq with the objectives of other producers. Iraq had sought to exercise sovereignty over its oil resource and remain independent of the marketing and technical expertise of the international companies. The governments of the large producers of the Persian Gulf and Libya, on the other hand, sought to maximize the revenues obtained from their oil under existing economic arrangements.[47] The seventh OPEC conference finally voted to refer acceptance or rejection of the royalty issue to the individual member countries.

OPEC's performance in the negotiations on royalty-expensing reflected the inability of producing governments to formulate a joint strategy or to cooperate on tactics. While negotiations were in progress, the countries discussed possible sanctions. These sanctions were to have taken the form of restrictions on production and various unilateral tax increases.[48] OPEC threats to resort to such measures in the absence of progress on the negotiations were not particularly credible because the organization had never implemented its threats. In November 1962, for example, OPEC announced that its members were unanimous in emphasizing the need for negotiations to be completed by 30 March 1963, and in the case of company noncompliance, it would consult to take steps to implement its resolutions on royalty-expensing.[49] The deadline passed but was ignored. One author has taken the position that OPEC's concentration on royalty and marketing expenses, as opposed to making demands for higher posted prices, was an error. He contends that the resolution of the royalty-expensing dispute left no issue which could rally producer support.[50]

---

[47]*Ibid.*, pp. 98-101. Stork makes much the same point about differing objectives among the producing governments, if with considerably more rhetorical flourish. Stork claims that Iraq alone attempted to assert sovereignty over its oil and that, guided by the Gulf states, OPEC in the 1960's pursued "diversionary objectives with desultory tactics" (*ibid.*, p. 102). As Mabro has pointed out, the ideological color of the regime generally has little effect on the objective of maximizing revenue. In fact, one can make the case that both "conservative" and "radical" governments have had very similar objectives (R.E. Mabro, "Libya," in *A Survey of North West Africa*, ed. W. Knapp [Oxford: Oxford University Press, 1976]).

[48]Stocking, *Middle East Oil*, p. 366.

[49]"OPEC and the Oil Companies," Supplement to *Middle East Economic Survey* (hereinafter cited as *MEES*), 28 August 1964, p. 3.

[50]Alnasrawi, "Collective Bargaining Power in OPEC," p. 192.

ECONOMIC PARAMETERS OF THE BARGAINING MODEL

The apparent weaknesses in OPEC were symptomatic of a more fundamental problem. In a market in which there was a surfeit of oil at going prices, it was difficult for producing governments to take joint action on production controls. Given the inability of the producing governments to coordinate possible sanctions, the concentration of OPEC on royalty-expensing was a highly successful strategy because it involved the common interests of both the majors and the main producing countries.

The Iranian government was the first to sign an agreement with the consortium on royalty-expensing; by 25 January 1965, Saudi Arabia and Qatar had also signed. It was in the interest of the majors to harmonize tax-paid costs in the main producing areas as soon as possible. Therefore, they stipulated that if any four countries had signed agreements by 26 January 1965, the royalty-expensing provisions would be made retroactive to 1 January 1964. Since Kuwait's finance minister had approved an agreement before the stipulated date,[51] the majors made retroactive payments to Saudi Arabia, Iran, and Qatar.

After these agreements had been negotiated, Libya became a battleground between the independent and major companies. When the Libyan government announced that it was planning to accept a royalty-expensing agreement on the same terms as those signed by the Persian Gulf countries, the independents protested directly to the government.[52] A group of representatives from the independent companies attempted to persuade the Libyan prime minister that lower government per-barrel revenues and a higher volume of production would both serve the interests of the independent companies and maximize Libyan revenue. They claimed that with higher per-barrel taxes the majors would "drastically reduce Libyan production by substituting Arabian Gulf oil,"[53] and they pointed to

---

[51] Approval of an agreement by Kuwait was delayed until 1967 because it became an issue for the Arab nationalist bloc in the newly formed National Assembly. The terms of the final agreement (made retroactive to 1966) were generally similar to the original royalty-expensing agreement negotiated with Iran. Thus domestic party politics seems only to have delayed implementation of the basic strategy of the Kuwait finance minister which had been devised in cooperation with other OPEC governments.

[52] Wanda M. Jablonski, "Libya's Oil Pricing and Tax Dilemma," *PIW*, 19 April 1965, p. 65.

[53] *PIW*, 1 November 1965, p. 8.

the low rates of expansion of production in Abu Dhabi[54] and Oman, where the concessions were held by the majors.

The most important factor in determining the outcome of the conflict in Libya between the independents and the majors was the Libyan government's desire for greater revenue. When the independents did not meet the requirements of a November 1965 decree relating to royalty-expensing, the prime minister threatened to halt immediately the exports of companies which did not comply. This threat was supported by legislation specifically empowering the council of ministers to terminate production and expropriate company assets. The independents had no other source of low-cost crude, and by January 1966 all of them capitulated to the new fiscal terms.

Although the majors won the battle over royalty-expensing in Libya, it soon became apparent that they had miscalculated the effects of equalizing tax-paid costs. When the independent companies in Libya were compelled to pay about 35 cents/bbl. more in government revenue, expectations were widespread that market prices would stabilize everywhere.[55] Prices stabilized only briefly, however, and then continued their decline (see Figure 1 above). The historic structure of prices had been undermined by the expanding market share of the independents and by the growing market in third-party transactions, including crude sales by Communist countries. Price leadership by the majors became less effective,[56] and the majors' attempt to use OPEC to harmonize tax-paid costs for a short-term advantage had long-term costs.

In the negotiations concerning royalty-expensing, the companies were forced to recognize the ability of producing countries to bargain collectively.[57] (The previous structure of negotiations—a consortium of companies with producing interests in more than one country facing a single producing government—had given the

[54]On 2 December 1971, Abu Dhabi established the United Arab Emirates in combination with the six other Trucial states: Dubai, Ajman, Fujairah, Sharjah, Umm al Qaiwan, and Ras al Khaimah.

[55]Adelman, *World Petroleum Market*, p. 189.

[56]Prices of petroleum products in Western Europe during 1960-69 were increasingly influenced by the markets for "incremental crude" in Rotterdam and Italy. At the light end of the barrel, gasoline was the focus of price competition led by the independents; at the heavy end of the barrel, fuel oil prices were cut by the majors. The only rise in bulk product prices in Europe occurred as a result of the Arab-Israeli war (Christopher Tugendhat and Adrian Hamilton, *Oil: The Biggest Business* [London: Eyre Methuen, 1968], p. 170).

[57]"OPEC and the Oil Companies," Supplement to *MEES*, 28 August 1964, p. 11.

companies considerably more leverage.) Moreover, the consultations among OPEC governments during the royalty-expensing negotiations made future cooperative efforts likely. Finally, a number of damaging precedents had been established in Libya during the final phase of the royalty-expensing battle. For one thing, the Libyan government had unilaterally changed the terms of a concession by royal decree. More important, the quick capitulation of the independents had indicated the power even one producing government had when negotiating with operating companies without diversified sources of crude. This was one of the key factors in the destruction of the system of price determination by bilateral bargaining which had characterized the 1960-69 period.

A new round of negotiations on royalty-expensing opened in 1967. In September 1967, the government of Saudi Arabia and the owners of Aramco reached an agreement whereby the exports of Saudi Arabia from Mediterranean terminals would be taxed without the 6.5 percent discount from the posted price; a similar agreement was soon reached with Libya.[58] The blockage of the Suez canal in 1967 enhanced the freight advantage for crude shipped from Mediterranean terminals, and in the absence of such an adjustment, increased production from Iraq, Libya, and Saudi Arabia would have exerted further downward pressure on market prices.

D. CONCLUSIONS

We now can state our basic conclusions concerning (a) the economic characteristics of the market for Middle Eastern crude and (b) the process of price determination during the 1960-69 period.

Our first basic conclusion is that the economic costs of crude oil production in the Middle East during 1960-69 were extremely low, and the historic pattern of concessions which had given the major companies joint control over the richest fields was a controlling factor in explaining the price structure for crude. Through joint producing agreements, each company was informed about the production and marketing plans of the others. The majors were thus able to maintain market shares jointly, to restrict the expansion of productive capacity, and to realize a monopoly rent on their exports

---

[58]The agreement also provided for a new schedule for the phasing-out by 1971 of the discounts from posted prices; as a result, total OPEC revenues increased by $125 million in 1968. (See *MEES*, 12 January 1968, p. 9.)

of crude oil.[59] They were also able to "adjust" supply to demand at a higher-than-competitive price level because they were vertically integrated.[60] Vertical integration allowed the majors to extend their control from crude markets to product markets, and to shift competition into such areas as marketing, where the structure of crude prices would not be threatened.[61]

The majors' control over production and price was significantly eroded during the 1960's. The independent companies obtained access to low-cost Middle East and North African production; this access, together with Communist bloc crude sales, expanded supply, and market prices fell. The majors and independents had largely incompatible objectives: the majors wanted to adjust supply to going prices in order to maintain their margins per-barrel while the independents were willing to cut prices in order to expand their market shares.

Our second basic conclusion is that crude oil prices during 1960-69 were the outcome of a process of bargaining among actors as outlined in Chapter I above.[62] In economic terms, the negotiations over per-barrel taxes were a struggle between companies and governments over the sharing of monopoly rent.[63] In the royalty-expensing negotiations, the majors perceived that limiting the expansion of supply would increase the shares of monopoly rent available both to companies and producing governments. The strategy was to pay higher per-barrel revenues if the producers would harmonize tax-paid costs, and thereby prevent the independents from using Libyan tax provisions to expand supply and gain a larger share of the market. The trends in company profits per-barrel and in government revenues per-barrel (see Figure 2 above) demonstrate that the majors were not successful. Company per-barrel

---

[59] Joint ventures provided a legal means by which the major companies could restrict the expansion of productive capacity (Testimony of John M. Blair, 25 July 1974, in U.S. Congress, Senate, Committee on Foreign Relations, Subcommittee on Multinational Corporations, *Multinational Corporations and U.S. Foreign Policy*, Hearings, 93rd Cong., 1st and 2nd sess. [hereinafter cited as *MNC Hearings*], 1975, 9: 195, 210-11, 218-19).

[60] Edith T. Penrose, *The International Oil Industry in the Middle East* (Cairo: National Bank of Egypt, 1968), pp. 10-11.

[61] Helmut J. Frank, *Crude Oil Prices in the Middle East: A Study in Oligopolistic Price Behavior* (New York: Praeger, 1966), pp. 132-33.

[62] See pp. 8-10.

[63] Michael V. Posner, *Fuel Policy: A Study in Applied Economics* (London: Macmillan, 1973), p. 52.

margins fell, and producers' revenues constituted most of the cost of a barrel of oil by 1970. Also, Libyan production continued to expand rapidly after the adoption of royalty-expensing. According to one company executive, by 1970 the companies were using their market power to act as "a commodity stabilizing agency for the crude oil producing countries."[64]

Our third basic conclusion is that the producing governments' ability to raise revenues shows a shift of bargaining power to their side through the mechanism of collaboration and collective bargaining. At the end of the decade, the rough balance of power between companies and governments had ended. Since bargaining during 1960-69 was largely a process of issuing threats and manipulating the credibility of such threats, it was not apparent to what extent the distribution of power had actually shifted. That became clear only when a producing government took the final step of imposing sanctions.

---

[64] Chandler, "Myth of Oil Power," p. 713.

PART TWO

THE MARKET IN DISEQUILIBRIUM

Chapter III

THE LIBYAN NEGOTIATIONS OF 1970

James Akins has observed that the Libyan demands of 1970 and their success demonstrated—"like a flash of lightning in a summer sky"—a fundamentally new situation in the oil market.[1] The policies adopted by the Libyan Revolutionary Command Council (RCC), which had seized power in Libya on 1 September 1969, proved decisive in upsetting the balance of power between OPEC and the oil companies. The tactics of the RCC revealed the vulnerability of the companies and of the consuming countries, and therefore had a profound impact upon the future policies of other OPEC members.

The challenge to the established pattern of negotiations on crude oil prices was mounted by a small group of officers with absolute power who were motivated by a semi-religious sense of nationalism. The RCC succeeded because it took the step of imposing production cuts, and because its threats of nationalization were highly credible. It is unlikely that any other producing government would have entertained such a course of action. The example of Iran under Mossadegh—where the loss of oil revenues following the nationalization of the oil industry in 1951 completely disrupted the nation's economy—clearly demonstrated the danger of implementing threats of nationalization.[2] If one considers the Iranian example too remote, there was the abortive 1967 Arab oil embargo against the West which was followed by production cuts that generally hurt the Arab nations.[3] The Libyan negotiations of 1970 were a watershed in the development of the international oil market. The outcome of these negotiations was determined primarily by two factors: (a) a short-term tanker shortage and (b) the bringing together of two sets of

[1] James E. Akins, "The Oil Crisis: This Time the Wolf Is Here," *Foreign Affairs* 51 (April 1973): 471.

[2] For a detailed account of the effects of the Iranian nationalization, see Benjamin Shwadran, *The Middle East, Oil and the Great Powers* (New York: Wiley, 1973), pp. 89-123.

[3] M. A. Adelman, "Is the Oil Shortage Real? Oil Companies as OPEC Tax Collectors," *Foreign Policy* 9 (Winter 1972-73): 90.

actors who were willing to break the "rules of the game" in negotiations over price. On one side was a determined military dictatorship, and on the other side, independent oil companies without alternative sources of supply.

A. OIL PRICES AND LIBYAN POLITICAL OBJECTIVES

King Idris of Libya had demanded that negotiations on the price of Libyan crude open in September 1969—the very month during which the king was overthrown by the RCC after a reign of nearly eighteen years. The change in government did not interrupt the flow of Libyan oil: the United States, Britain, and France were notified by the new regime that it would respect all commercial agreements then in force.[4] In fact, for nine months after the revolution, oil production continued to climb, reaching a peak of 3.7 million b/d in April 1970.[5]

There was little in the initial policies of the RCC to distinguish it from the classic coup d'etat. The major powers in the region—Egypt, Britain, and the United States—knew almost nothing about the leadership of the new government during the early days of September 1969, which allowed the RCC to consolidate power. Shortly thereafter the characteristics of the new regime became evident: churches were closed, foreign privileges were revoked, royal projects were suspended, and a ban was placed on alcohol.[6] There was an emphasis on the symbols of independence and a return to Islamic tenets.

The oil companies operating in Libya remained largely unaffected by the legislation of the new government, but company executives were anticipating demands for higher revenues. For example, an Oasis internal memorandum circulated in September-October 1969 stated:

> Once the regime is stable, it will launch a frontal attack on the oil industry. Driven by missionary zeal of secure absolute economic sovereignty, the regime will use every possible means of 'persua-

---

[4] First, *Libya*, p. 99.

[5] *Petroleum Press Service* (hereinafter cited as *PPS*), June 1970, p. 221.

[6] W. B. Fisher, "Libya: Physical and Social Geography," in *The Middle East and North Africa 1975-1976*, p. 529.

sion' but it is unlikely that it will resort to outright expropriation.[7]

Why did the militant RCC refrain from an immediate assault on the companies? Breton suggests that Prime Minister Mu'ammar Gadafi rejected a course of action that would threaten oil revenues, which he wanted to use to promote Arab cooperation and to aid "in the battle to liberate Palestine."[8] Also, there are indications that delay was required by the new regime to give it time to plan its strategy.[9]

During the first part of 1970, the RCC articulated four objectives to be achieved in its negotiations with the operating companies: (a) to increase substantially government per-barrel revenue, (b) to increase government control over producing operations, (c) to demonstrate that "Arab socialism is the wave of the future," and (d) to utilize increased oil revenues in aid of the "battle to liberate Palestine." To these goals a fifth could be added—namely, to discredit the previous regime by forcing the companies to admit that Libyan crude had been underpriced.[10]

B. THE MANIPULATION OF THREATS

The Libyan oil minister opened discussions on posted prices with representatives of twenty-one operating companies on 20 January 1971. Prime Minister Gadafi addressed the group and made the widely reported statement that "people who have lived for 5000 years without petroleum are able to live without it even for scores of years in order to reach their legitimate rights."[11] The government's

---

[7] Cited in First, *Libya*, p. 200. (Oasis was a consortium of three American independent companies and Shell.)

[8] Hubert Breton, "Le pétrole libyen au service de l'unité arabe?," *Revue Francaise de Science Politique* 22 (décembre 1972): 1260-61.

[9] Another explanation for the delay is an alleged conflict over tactics between Deputy Prime Minister Abdessalam Jallound and Gadafi. Jallound is reported to have persuaded Gadafi to abandon plans for nationalization in favor of demands for higher taxes (see Leonard Mosley, *Power Play: Oil in the Middle East* [Baltimore: Penguin, 1973], pp. 354-57).

[10] Statement by George Henry Mayer Schuler, Chief London Policy Group Representative for the Bunker Hunt Oil Co., in *MNC Hearings* 6: 2-3. [This statement was reprinted as G. H. M. Schuler, "The International Oil 'Debacle' Since 1971," Special Supplement to *PIW*, 22 April 1974, pp. 1-36.]

[11] U.S. Library of Congress, Congressional Research Service, Foreign Affairs

objective, according to the industry press, was an increase of 40-45 cents/bbl. in posted prices.

About one-third of Libyan crude liftings during the first half of 1970 represented production by independent companies—a larger share than in any other major exporting country.[12] Mobil, BP, Shell, Texaco, Esso, and Socal—six of the seven majors—lifted only a small proportion of their total crude in Libya. These companies were very concerned that an increase in posted prices in Libya might result in the "leapfrogging" of similar demands into the Persian Gulf countries which supplied most of their crude. Thus the majors had every reason to be militant in refusing higher posted prices.[13] On the other hand, for independent companies higher taxes were a better alternative than being deprived of an assured low-cost source of production and having to turn to the third-party market.

After the initial meeting between Libyan officials and the operating companies, the Libyan negotiating team began separate talks with Occidental and Esso. The selection of these two companies was obvious: Occidental had the largest proven reserves and Esso was the largest single producer. The Libyan negotiating position was based upon its charge that Libyan crude had been underposted since 1961, and that since 1967 posted prices had not reflected the increased freight advantage of Libyan crudes in light of the closing of the Suez Canal.

The Libyan demands threatened company profitability in a number of respects: (a) if posted prices were increased, royalties would also increase, and royalty payments could not be credited against U. S. taxes; (b) while some concession could be made for the transport premium that Libyan oil commanded, a large increase

Division, *Chronology of the Libyan Oil Negotiations, 1970-1971*, prepared for the Subcommittee on Multinational Corporations, Committee on Foreign Affairs, U. S. Senate, 93d Cong., 2d sess., 1972, p. 3.

[12] Libyan production was approximately 3.5 million b/d for the first half of 1970. Oasis was the largest producer at one million b/d; Occidental was producing at 780,000 b/d; an Esso subsidiary was producing at 750,000 b/d; BP-Hunt (a consortium of BP and Bunker Hunt) at 400,000 b/d; Amoseas (a consortium of Texaco and Socal) at 360,000 b/d; and Mobil-Gelsenberg (of which the German state-supported company had a one-quarter share) was producing at 270,000 b/d (U. S. Department of the Interior, Bureau of Mines, *Minerals Yearbook 1970*, House Document no. 92-22, 1972, 3: 515).

[13] George Piercy, Director of Exxon, testified on 1 February 1974 that it was clear at the time of the negotiations that Libyan demands were "something that would establish a new mark and would have spillover possibilities into the Gulf" (*MNC Hearings* 5: 196).

in posted prices would undermine the entire fiscal structure in the Persian Gulf; (c) the increase in the tax rate applied to posted prices demanded by the Libyan negotiators to compensate for alleged past "underposting" would violate the 50-50 principle of "profit sharing" which had been in effect since 1950; and (d) acceptance by the companies of retroactive tax increases would be a dangerous precedent and might upset the existing pattern of negotiations.

The negotiations with Occidental and Esso made little progress, and were suspended on 22 April 1970 when the Libyan government rejected the companies' offer of a 6 cents/bbl. increase in posted price.[14] During the negotiations the Libyan leaders issued a number of threats. For example, in early April 1970 Gadafi threatened unilateral action if the companies remained intransigent. A few days later, the Libyan oil minister hinted at the possibility of nationalization. More significant than such threats, however, was a joint communique signed by Algeria, Libya, and Iraq in which the three major Mediterranean producers pledged mutual support in negotiations with the companies.

Up to this point, the escalating threats and increasing amounts offered by the companies followed the previous pattern of bargaining. The issue was higher government revenue, although it was couched by the Libyans in terms of their "legitimate rights." As late as May 1970, informed opinion was that a compromise would be reached. However, the Libyan government refused to moderate its demands; it adopted a new strategy, and escalated the conflict by imposing sanctions upon the operating companies.

C. STRATEGY AND ECONOMIC SANCTIONS

Negotiations with Occidental and Esso were renewed in early May, but were again suspended at the end of May, and in June the Libyan government imposed new port dues. The companies were not permitted to claim additional production expenses in their discounts from posted price, so higher port dues were tantamount to an increase in government revenue per-barrel. Late in May, the Libyans took the additional step of imposing limitations on Occidental's production, which reduced it from a peak of 800,000 b/d in April to 485,000 b/d in June. The government simultaneously refused to permit Esso to export liquified gas from its $350 million facility. On 15 June the Libyan government required Amoseas to cut produc-

---

[14] Tugendhat and Hamilton, *Oil*, p. 182.

tion by 31 percent to 275,000 b/d; on 20 July Oasis was required to cut production by 12 percent to 895,000 b/d; on 15 August Mobil-Gelsenberg was required to cut production by 20 percent to 222,000 b/d; and on 5 September Esso was required to cut production about 15 percent to 630,000 b/d.[15] In July, Libya nationalized (with compensation) the distribution facilities of the three main oil marketing companies—Shell, Esso, and an ENI subsidiary. Nationalization was widely interpreted as an indication of the government's determination to take further steps if its demands for higher revenues were not met.[16]

The most important sanctions imposed by the Libyan government were the progressive cutbacks in production. The government denied that such measures were related to the price negotiations, and it justified the new ceilings on production in terms of the conservation of limited oil resources. It is true that there were charges during the late 1960's that some of the companies were depleting Libyan fields faster than the maximum efficient rate,[17] and there is evidence to indicate that Occidental, in particular, was producing at an excessive rate—which might be expected from a firm attempting to maximize profits under conditions of political uncertainty. However, even Occidental was not overproducing by an amount approaching the size of the imposed cutbacks.[18] The chief purpose of the production limitations was clearly to force the companies to accept demands for higher posted prices.[19]

The Libyan strategy was directed at the independent companies. The choice of Occidental for special pressure was astute: Occidental had no other source of foreign crude;[20] it depended upon

---

[15] Statement by G. H. M. Schuler, in *MNC Hearings* 6: 3.

[16] Taki Rifaï, "La crise pétrolière internationale (1970-1971): Essai d'interprétation," *Revue Francaise de Science Politique* 22 (décembre 1972): 1199. Much of the material in this article was published later in Taki Rifaï, *The Pricing of Crude Oil: Economic and Strategic Guidelines for an International Energy Policy* (New York: Praeger, 1974).

[17] Every oil field has a so-called maximum efficient rate determined by geological and technological factors; if this rate is exceeded, the total amount of oil-in-place ultimately recoverable is reduced.

[18] Testimony of James E. Akins, 11 October 1973, in *MNC Hearings* 5: 10.

[19] When Occidental accepted higher posted prices in September 1970, it was again authorized to produce at 700,000 b/d.

[20] Both Occidental and the Oasis consortium (with the exception of Shell) had no alternative sources of crude (see Testimony of James E. Akins, in *MNC Hearings* 5: 16).

its Libyan operations for almost one-third of its revenues; and given its contractual commitments, it could not survive a drastic cut in production.[21]

The timing of the Libyan move also maximized its effectiveness. The Suez canal had been closed since 1967, and the Tapline[22] had been cut in Syria on 3 May 1970, with the Syrian government refusing to allow repairs until it received higher transit duties.[23] By August 1970, the production of the companies in Libya had been reduced by almost 800,000 b/d, and the Tapline cut meant that an additional 500,000 b/d of Saudi crude could not be moved to the Mediterranean port of Sidon in Lebanon. Although the total amount of crude in shortfall was not large—a maximum of about 1.3 million b/d, for which there was more than adequate spare capacity in the Gulf[24]—the shortfall was in short-haul Mediterranean crude, and the replacement of this production with Gulf crudes required a fourfold increase in tanker tonnage from the Gulf to European ports.[25]

1. Bargaining Strength and Supply Logistics. The market in tanker charters closely approximates the conditions of pure competition,[26] but it remains imperfectly competitive. Investors do not have perfect knowledge of future demands, and since 1970 there has been a two- to three-year lag between the order and the delivery of a tankship. Therefore, the spot tanker rate is subject to great fluctuation in response to shifts in demand, even though the long-term freight rate is set by the cost of supplying a marginal ship.

In 1969-70 the demand for oil products increased more rapidly than expected in Western Europe,[27] and by May 1970, reflecting an

---

[21] Occidental's crude sales contracts had penalties for non-delivery, but enabled the company to shift increased taxes partly or substantially to buyers (Mikdashi, *Community of Oil-Exporting Countries*, p. 147).

[22] The Tapline runs for a distance of 1,040 miles from the Saudi Arabian oil fields through Syria to the coast of Lebanon.

[23] Repairs were made late in January 1971, after the operating company doubled annual payments to the Syrian government to $8 million.

[24] The mid-July shortfall was calculated at 1.07 million b/d (see *MEES*, 17 July 1970, p. 5).

[25] *PPS*, October 1970, p. 398.

[26] Adelman, *World Petroleum Market*, p. 109.

[27] The percentage increase in consumption in Western Europe for 1969-70 was 11.4 percent, compared to an annual average increase from 1965-70 of 10.0 percent (*BP Statistical Review 1970*, p. 8).

already chronic shortage of tankers, the spot tanker rate was the highest it had been since the oil embargo of late 1967. Prices for petroleum products also rose, and company earnings increased. The Libyan negotiators claimed that there was an inadequate premium for the freight advantage of Libyan crudes. The cutbacks imposed by the RCC in Libya enhanced the freight advantage of Libyan crudes because additional crude was available only from the more distant Persian Gulf;[28] the rate for a single tanker voyage from the Persian Gulf to Rotterdam increased from $1.10/bbl. in late May to $3.00/bbl. in September 1970.[29]

The characteristics of the tanker market had crucial implications for the process of bargaining over per-barrel taxes. Tanker rates were sure to fluctuate in a competitive market and give rise to substantial differentials in the market value of North African and Persian Gulf crudes. Producing governments were sensitive to the fact that company margins were temporarily higher, and renewed their demands for higher per-barrel taxes. However, when higher taxes were negotiated by a producing government because of freight advantages, it triggered demands for revenue increases by other producing governments even though the market price of crude in those other countries had fallen because of the higher freight rates.[30]

Another implication of increases in tanker rates for negotiations over price was that such increases severely undermined the bargaining strength of the majors. The majors could not neutralize the production ceilings imposed in Libya by shifting production to other sources of supply.[31] Either the tanker tonnage was not available to haul crude from the Persian Gulf, or it was available at a price far in

---

[28] The combined effect of the Tapline closure and the Libyan cutbacks was to increase the total demand for tanker tonnage by 6-7 percent (Testimony of George Piercy, in *MNC Hearings* 5: 196).

[29] *PE*, December 1970, p. 475.

[30] Stocking notes that OPEC bargaining in the 1960's, except for duration, did "not differ radically from industry-wide industrial union bargaining" (*Middle East Oil*, p. 361). There are important parallels between bargaining over oil prices and bargaining over wages. For example, to take a simple model, assume that one union represents workers in an industry which has high profits due to high labor productivity. That union is sensitive to higher profits and demands an increase in wages. The higher wage settlement causes other unions to raise *their* wage demands, despite the fact that labor productivity in other industries may be decreasing. In this model, higher profits in any one unionized industry will produce a new round of wage demands in other sectors of the economy.

[31] Mabro, "Libya."

THE LIBYAN NEGOTIATIONS OF 1970

excess of the Libyan demands for per-barrel tax increases. Because of the tanker shortage, Libya was successful in imposing production cutbacks and did not need to coordinate its action with other producers.

The significance of the tanker shortage is often missed, and the whole course of events misinterpreted. Most commentators are of the opinion that the producing countries obtained higher revenues because of a surge in demand which turned a buyers' market into a sellers' market.[32] This theory bears no resemblance to the facts: the 1970 increase in consumption over 1969 was somewhat below the 1970-71 average in all areas; the increase in 1971 over 1970 in Western Europe and Japan was about half the decade's average;[33] by 1972 there was substantial excess producing capacity. Thus Libya's tactical opportunity was the existence of a short-term shortage in tankers. In the year 1970 there may have been a transportation crisis, but there was certainly no surge in demand that met limited supply.[34] Not only was there an excess of productive capacity in the Persian Gulf, but other producing governments were willing—but unable—to use Libya's cutbacks as an opportunity to increase their market shares.[35]

2. International Cooperation and Bargaining Strength. We should point out that there is a fundamental difference between the joint action of nations on productions controls and declarations of solidarity in forthcoming negotiations. The governments of Libya, Algeria, and Iraq only consulted with each other during the Libyan negotiations of 1970. In addition, Algeria's special relationship with France put it outside the general Western market, but when Algeria unilaterally raised its posted price for exports to France in late July

---

[32] For example: "The situation changed rather abruptly in 1970-71, when demand growth finally overtook capacity. . . ." (James W. McKie, "The Political Economy of World Petroleum," *American Economic Review* 64 [May 1974]: 51).

[33] Adelman, "Is the Oil Shortage Real?," p. 72.

[34] J. E. Hartshorn, "Oil Diplomacy: The New Approach," *The World Today* 19 (July 1973): 285. See also Hartshorn, "From Tripoli to Teheran," pp. 291-92.

[35] As late as 1971, the Iranian prime minister announced that the principal concern of his government was to expand oil exports by 20 percent. At this time, Saudi Arabia continued to exert pressure on Aramco to double production within five years (Rifai, "La crise pétrolière," p. 1210). Mikdashi reports similar "inside" information in *Community of Oil-Exporting Countries*, p. 150.

1970, this enhanced the credibility of Libyan threats. Libya, however, was the only Mediterranean exporter to impose production ceilings. Although the Libyan government's demands for tax increases were coordinated with the Algerian government, the imposition of sanctions by Libya was not.[36] Libya's bargaining strength was not a function of cooperative action with other producers; rather, it reflected Libya's adroit use of tactics to successfully exploit a short-term advantage.[37]

Libya's bargaining strength was enhanced by its substantial foreign exchange holdings.[38] Since oil revenues accounted for more than 60 percent of the Libyan GNP, and since a great deal of employment in Libya was a mechanism for distributing oil wealth,[39] the imposition of production cuts without financial reserves would have disrupted the country's economic life. The fact that the majors had previously agreed to higher government revenues had increased Libya's financial reserves and thus strengthened the negotiating position of the new government.

The Libyan tactic of holding separate negotiations with each operating company was designed to make it more difficult for the companies to act together. In July 1970, Occidental's situation was becoming increasingly precarious. A. Hammer, the chief executive of Occidental, approached J. K. Jamieson, the chief executive of Esso, for the purpose of obtaining alternative sources of oil. Jamieson reportedly agreed to guarantee to provide Occidental oil at going third-party rates. Hammer, however, wanted oil at close to tax-paid cost, and no agreement was reached.[40] Esso thus rejected an early attempt to devise a safety-net agreement that could thwart the Libyan strategy of applying pressure to the most vulnerable indepen-

---

[36] The Algerian increase in posted price corresponded with the goal of Libya in negotiations (Statement of G. H. M. Schuler, in *MNC Hearings* 6: 2).

[37] See Hartshorn, "From Tripoli to Teheran," p. 292.

[38] Libya's gross reserves were $1,590 million at the end of 1970 (International Monetary Fund, *International Financial Statistics,* June 1971, p. 18). This would have been enough to cover government expenditures and imports at then current levels for four years (see Akins, "Oil Crisis," p. 471).

[39] On this point, see Robert Mabro, "La Libye: un état rentier?" *Projet* 39 (novembre 1969): 1090-1101.

[40] See Anthony Sampson, *The Seven Sisters: The Great Oil Companies and the World They Made* (London: Hodder and Stoughton, 1975), p. 212. Sampson's account is based upon an interview with Jamieson, and is in agreement with the information obtained by the staff of the Senate Subcommittee on Multinational Corporations (see *MNC Hearings* 5: 108).

THE LIBYAN NEGOTIATIONS OF 1970

dents. In August 1970, Esso made an offer of a 10 cents/bbl. increase in government revenue, and a supplemental 11 cents/bbl. as a freight premium which would fluctuate with the spot tanker rate.[41] This was the first attempt to deal with the problem of tankship rate fluctuations. The Libyans, however, rejected the offer.

On 4 September 1970 Gadafi announced the signing of an agreement with Occidental. The Occidental settlement was not a compromise, but rather a complete capitulation to the Libyan demands.[42] The terms of the agreement with Occidental involved an increase of 30 cents/bbl. in posted price for 40° API crude. Postings were also to escalate at two cents per year for the next five years. The most important aspect of the agreement was an increase in the government's tax rate from 50 to 58 percent—violating the so-called 50-50 principle. The justification for the violation of 50-50 profit sharing was that companies were to make tax payments retroactive to 1965 in order to compensate for the underposting of Libyan crude since that date, and the government required explicit written acknowledgment of that fact. The net result of the new fiscal provisions was that the government "take" rose by about 27 cents/bbl.

The operating companies now required a high degree of cooperation if they were to maintain a common negotiating front. Six days after the settlement with Occidental, the other producers of Libyan crude met in New York City to outline possible strategies for resisting similar price and tax increases. Concerned with possible violations of American antitrust legislation and with possible further sanctions by the Libyan government, Occidental refused to supply details of its agreement to the other producers.[43] Having failed in a second attempt to work out a program which would strengthen the ability of individual companies to resist the Libyan sanctions, the other independents in late September settled with the RCC on terms similar to those accepted by Occidental. The new Libyan tax rates were approximately 55 percent, and the maximum increase in government "take" was 32 cents/bbl. The majors were now concerned that the Libyans would unilaterally increase posted prices and tax rates; therefore, except for Shell, they settled with the Libyan government in early October. Shell did not settle until December, but its stakes in Libyan production were minimal. It took the position that the principle of retroactive payments could justify any kind of producing-

---

[41] Testimony of George Piercy, in *MNC Hearings* 5: 197.

[42] Testimony of James E. Akins, in *ibid.* 5: 19.

[43] Statement of G. H. M. Schuler, in *ibid.* 6: 4.

government claim. Shell's output share in the Oasis fields (150,000 b/d) was suspended from September until its settlement in December.

3. Consuming-Country Intervention and Bargaining Strength. In the Libyan negotiations of 1970 the major actors remained the oil companies and the producing governments. As in the 1960's, the role of the governments of the countries that represented consumers of oil products could be characterized fairly simply: "They stayed out of negotiations and paid up to the winner."[44] Consuming-country nonintervention in price negotiations (France and Italy were exceptions)[45] was not entirely unsuccessful: large amounts of oil were available at a price well below other sources of energy. The difference between supply cost and market price encouraged companies to engage in extensive exploration activities and led to discoveries of large reserves.[46] It was advantageous to the consuming countries to have the companies negotiate on price because such negotiations were considered commercial disputes rather than political confrontations. Traditionally, the major companies have seen themselves as a necessary commercial buffer or as "the reconcilers of the objectives of producer and consumer countries."[47] This has been particularly important in relations between the U. S. and Arab oil suppliers, because of U. S. military and political support of Israel.

The U. S. Department of State played a rather marginal role in the Libyan negotiations of 1970, and there has been considerable criticism of the Department for not taking action to counter the Libyan demands.[48] What were the reasons for such inaction? Un-

---

[44] Hartshorn, "From Tripoli to Teheran," p. 291.

[45] See p. 117 below.

[46] The trade press is wont to point this out (see, for example, *PE*, May 1975, p. 162). Company executives sometimes refer to the 1960-70 period as "the golden age."

[47] George Glass, "The Changing Dimensions of Government-Company Relationships," Supplement to *MEES*, 17 May 1968, p. 3. (Glass was the head of trade relations for the Middle East at Shell.) See also Statement of William P. Tavourlareas, President of Mobil Oil, 6 June 1974, in *MNC Hearings* 9: 81-86. Tavourlareas argues that commercial relations are to be preferred to government-to-government negotiations on supply and price.

[48] Adelman criticizes the Department of State's handling of the Libyan negotiations, and contends that without U. S. support, OPEC would have never been able to achieve the solidarity necessary for further increases in price: "When the first Libyan cutbacks were decreed in May 1970, the U.S. could have

doubtedly one factor influencing the State Department's policy was the rapid growth of U. S. imports of Middle East oil. In 1970, after twelve years of continuous increases, U. S. domestic production began to decline. In the same year, the quotas imposed under the U. S. Mandatory Oil Import Program were successively relaxed in June, October, and December.[49] By the end of the year, it was already clear that the growth in U. S. imports (estimated by the President's Task Force on Oil Import Control in February) had been seriously understated.[50] State Department officials adopted the position that an increase in per-barrel revenues in Libya was justified by market conditions, and that a settlement on terms favorable to the Libyan government would make it less likely that oil supplies would be interrupted in the future.[51] Ensuring future security of supply had a higher priority than the maintenance of lower prices. (It might also be noted that the United States had a large protected domestic oil industry; for example, American wellhead prices for crude were $1.30/bbl. above world market prices.[52] Also, Western European consumers would pay most of the increase in taxes to the Libyan government.)

Shortly after Occidental's agreement with the RCC and the failure to work out a joint company strategy at the New York City meeting, the companies approached the U. S. government and urged government intervention in the negotiations. Sir David Barran, the chief executive officer of Shell, argued that the interests of both companies and consumers lay in resisting the Libyan demands. Capitulation to these demands, he argued, would mean having to face escalating demands from other producing governments.[53]

easily convened the oil companies to work out an insurance scheme whereby any single company forced to shut down would have crude oil supplied at tax-plus-cost from another source. . . . The OPEC nations were unprepared for conflict. Their unity would have been severely tested and probably destroyed. The revenue losses of Libya would have been gains to all other producing nations, and all would have realized the danger of trying to pressure the consuming countries" (Adelman, "Is the Oil Shortage Real?," pp. 79-80, and Adelman, *World Petroleum Market*, p. 284).

[49] Senate Committee on Finance, *World Oil Developments*, pp. 66-67.

[50] U. S. Cabinet Task Force on Oil Import Control, *The Oil Import Question: A Report on the Relationship of Oil Imports to the National Security* (Washington: GPO, February 1970).

[51] Testimony of James E. Akins, in *MNC Hearings* 5: 4-10.

[52] Estimated by the U.S. Cabinet Task Force on Oil Import Control, *Oil Import Question*, p. 405.

[53] U. S. Congress, Senate, Committee on Foreign Relations, Subcommittee

THE POLITICS OF CRUDE OIL PRICING

In 1970, Libya supplied almost one-quarter of the total European demand for oil, but Western European countries were "almost unaware of what was happening and would have been totally unprepared if oil had been cut off."[54] One State Department official reports that a top executive of a major oil company urged the American government to encourage the Libyan government to nationalize oil company producing assets because this step would finally awaken Western European governments to the seriousness of the situation.[55] The consuming countries had delegated bargaining on price to the companies, and consuming-country officials were ill-equipped to understand that Libya's success would seriously undermine the negotiating strength of the companies. Moreover, the consuming countries were divided on oil policy. One of the fears of the U. S. Department of State was the possibility that American companies might be recalcitrant and that, if there were a production shutdown, the European governments would make their own individual deals with the RCC.[56] In the end, the companies were informed by the Department of State that its representations would be of little help, and the companies operating in Libya that had not reached a settlement saw no other alternative than to meet the Libyan demands.[57]

D. CAPITULATION AND ITS CONSEQUENCES

Company executives had predicted that capitulation to the Libyan terms would encourage escalating demands by other producing governments, and such demands soon materialized. The Iraq Petroleum Company (IPC) unilaterally raised the posted price for Iraqi crude by 20 cents/bbl. in October 1970. The government of Iraq then demanded, and received, additional per-barrel payments that equalized their taxes with those of Libya. In negotiations with France, Algeria also insisted upon parity with the new Libyan posted prices.

on Multinational Corporations, *Report on Multinational Corporations and U. S. Foreign Policy*, 93d Cong., 2d sess., 1975, p. 125.

[54] Akins, "The Oil Crisis," p. 471.

[55] *Ibid.*, pp. 471-72.

[56] *Ibid.*, p. 472.

[57] Robert B. Krueger, *The United States and International Oil: A Report to the Federal Energy Administration on U.S. Firms and Government Policy* (New York: Praeger, 1975), p. 62.

THE LIBYAN NEGOTIATIONS OF 1970

The most important repercussions of the Libyan settlement were felt in the Persian Gulf. After threats by the Shah of Iran to expropriate assets of the operating companies, the consortium increased postings for Iranian crude by 9 cents/bbl. The 55 percent tax rate became the new standard, and the operating companies attempted to defuse further demands by immediately offering to apply this tax rate to Middle Eastern producers. Kuwait agreed on 14 November 1970 to a 9 cents/bbl. increase in posted price, but Aramco and Saudi Arabia were unable to reach an agreement.

In sum, the Libyan settlement marked the end of the pattern of bargaining on fiscal policy which had evolved during the decade of the 1960's. The policies of the major actors in the market explain the timing and degree of the 1970 price increase, but the shift in market power to the Libyan government also reflects a number of other factors. One factor was the heavy reliance of Western European countries on North African sources of supply. A second factor was the short-term tanker shortage which prevented the companies from shifting oil production to sources outside of Libya. A third factor was the rapid growth in the market share of the independent companies: the ability of the oil companies to play one producing government off against another now depended on the ability of oil company executives to reconcile the conflicting interests of majors and independents. The companies proved unequal to the task during the Libyan negotiations of 1970, and company appeals to the U. S. government to organize a joint company negotiating stance were also unsuccessful.

As we have pointed out, the Libyan success was not a product of greater cohesion among OPEC members. Rather, it is explained by the adroit tactics of the RCC when it perceived the advantages that a transportation crisis afforded. If the Libyan government had been confronted by the companies *and* the consumers, OPEC solidarity would have been dealt a major blow. Even if Libya had shut down production and market crude prices had risen dramatically, it is highly unlikely that in 1970 producing governments would have taken the joint action necessary to maintain an increase in per-barrel payments. The Libyan success, therefore, had profound implications for the future course of bargaining over price. Producing governments were encouraged to take immediate action to ensure that they would participate in any new tax increases. Most important, the Libyan example made future threats to shut down production more credible, and it demonstrated that the consuming governments were unable, or unwilling, to resist producing-government demands for higher taxes and prices.

Chapter IV

THE SHIFTING BALANCE OF MARKET POWER: THE ACCORDS OF TEHERAN AND TRIPOLI

Walter J. Levy observed in July 1971 that the balance of power among the oil companies, the oil-consuming countries, and the oil-producing countries had shifted decisively in favor of the oil producers.[1] This shift in the balance of power followed the negotiations of 1971-72 which substantially raised producing-governments' per-barrel revenues.[2] In this chapter we shall review the negotiations on oil prices during 1971-72, and analyze the policies that explain the growing power of the producing countries over oil production and price.

A. THE STRUCTURE OF THE NEGOTIATIONS

In 1971 OPEC states were far from a cohesive group of cooperating allies capable of regulating oil production. However, despite the hope retained by some company executives that the Persian Gulf states would exercise "restraint" in the aftermath of the Libyan settlement,[3] all producers—even the "conservative" Persian Gulf states—were determined to share in any tax increases which were granted to another producing country.

At the OPEC meeting held in Caracas from 9-12 December 1970, the members supported the terms of the Libyan settlement and demanded that higher posted prices and taxes be extended to all producers.[4] One commentator close to OPEC has observed: "The Libyan settlement was an embarrassment to other OPEC member

---

[1] "Oil Power," *Foreign Affairs* 49 (July 1971): 652.

[2] John Maffre, "Economic Report/Administration Searching for Methods to Aid U. S. Oil Interests in Bargaining with OPEC," *National Journal* 4 (15 May 1972): 808.

[3] Testimony of G. H. M. Schuler, in *MNC Hearings* 6: 4.

[4] *OPEC Resolution* XXI.120.

countries. It rendered further silence almost impossible."[5] The bargaining stance of the negotiators for the Persian Gulf states was also affected by recent actions of the Venezuelan government. Bolstered by high freight rates and the fact that its crude was sold at a premium in the American market because Venezuela was regarded as a secure source of supply, the Venezuelan Chamber of Deputies had passed an act in the beginning of December 1970 which increased its tax rate on the posted price of crude to a flat 60 percent and—more important—provided for unilateral determination of the tax reference price. During their subsequent negotiations with the companies operating in the Persian Gulf, government representatives threatened more than once to implement a "Venezuelan solution" unless their demands were met.

The Caracas meeting attempted to define the structure of producing-government negotiations with the operating companies. Renewed demands for higher revenues were justified by a general improvement in market outlook (although economic conditions remained basically unchanged except for the Libyan cutbacks).[6] At the Caracas meeting, a well-defined threat was issued to the operating companies: if its demands were not met, OPEC would respond with "concerted and simultaneous" action. The threat of a complete embargo did not have much credibility, however, since a basic objective of most of the Persian Gulf states continued to be an increase in the volume of production.[7] OPEC was careful to characterize its demands as a commercial disagreement with the operating companies rather than as a threat to impose an embargo on the consuming countries.[8]

The most important decision of the Caracas conference was to divide the producing governments into separate Persian Gulf and

---

[5] Abdul Amir Q. Kubbah, *OPEC: Past and Present* (Vienna: Petro-Economic Research Centre [1974]), p. 54.

[6] These demands were based upon OPEC Resolution XVI.90, which states that operating companies "shall not have the right to obtain excessively high net earnings after taxes, and that this shall be a cause for renegotiations."

[7] During the period of the negotiations, both productive capacity and the volume of exports in the Persian Gulf reached new highs (see Rifaï, "La crise pétrolière," p. 1210).

[8] Only the companies were threatened by a shutdown; presumably consuming governments would still be able to obtain oil at the higher prices (Kubbah, *OPEC*, p. 57). This shows the desire of the major producers to avoid having the negotiations characterized as a confrontation between producing and consuming governments.

Mediterranean groups for negotiations with the companies. This strategy was adopted by OPEC apparently because Libyan and Algerian demands related not only to higher revenues but also to "political issues," such as American policy in the Middle East. Conducting negotiations on a global basis, therefore, was felt to be potentially damaging to OPEC unity.[9]

The Libyan government was dissatisfied with the level of per-barrel increases demanded at the Caracas conference, and it refused to accept the OPEC decision that the Persian Gulf states should take the initiative in the negotiations.[10] Libyan officials met with representatives of the operating companies on 3 January 1971. The RCC made demands which they indicated were designed to coerce the companies to pressure the U. S. government to change its Middle Eastern policy.[11] The RCC also demanded measures which would have amounted to a 50 cents/bbl. increase in government revenue. Thus, three months after an agreement that was supposed to provide for a stable increase in government revenue over the next five years, the Libyan government made demands for additional taxes. This was an early indication that long-term promises of price stability had little value.

1. Company Cooperation and the Structure of Negotiations. Sir David Barran, chairman of Shell, was a key figure in the attempts to improve the negotiating strength of the oil industry by coordinating company action. Shell had been the holdout in the first round of the Libyan negotiations, and [the]

> Shell view was that the avalanche had begun and our best hope of withstanding the pressures being exerted by the members of OPEC would lie in the companies refusing to be picked off one by one in any country and by declining to deal with the producers except on a total global basis.[12]

Sir David suggested that companies operating in Libya should

---

[9]*Ibid.*

[10]Edith Penrose suggests that a basic Libyan objective was to establish Mediterranean oil as a "marker crude" in accordance with which taxes in other producing countries would be fixed ("The Development of Crisis," *Daedalus* 104 [Fall 1975]: 41).

[11]Statement by G. H. M. Schuler, in *MNC Hearings* 6: 7.

[12]Letter to Senator Frank Church from Sir David Barran, 16 August 1974, in *MNC Hearings* 8: 722.

meet and discuss a joint strategy, and such a meeting was held on 11 January 1971 in New York City in the law offices of John J. McCloy. U. S. Department of State and Department of Justice officials approved a joint company approach to the problem, and the Department of Justice stated that the companies would not be prosecuted for violations of American antitrust laws.[13] The meeting of company representatives resulted in three decisions: (a) the companies established an agreement to share crude oil among Libyan producers at close to tax-paid cost in the event that any company's Libyan production was cut back; (b) the operating companies in Libya—joined by the Compagnie Francaise des Pétroles (CFP), Petrofina of Belgium, Hispanoil of Spain, and the Arabian Oil Company of Japan—drafted a joint message to OPEC calling for negotiations on a global basis; and (c) the companies agreed to establish a London Policy Group (LPG) which would be responsible for coordinating the negotiating teams and authorizing final terms of settlement.[14] U. S. Undersecretary of State John N. Irwin was assigned to communicate the support of his government for a joint approach to negotiations to the heads of state of the Middle Eastern producing countries.

The oil-sharing agreement reached by the companies was an important, although belated, move to thwart the Libyan tactics of concentrating threats and sanctions on the most vulnerable independent companies. The companies' position on joint negotiations represented a dramatic change from their refusal to negotiate collectively with OPEC during the early 1960's. The global approach by the companies was an attempt to reach a settlement that would not precipitate another spiral of demands. The 16 January 1971 letter to OPEC members from the companies proposed a settlement which would make automatic adjustment in revenues for inflation and for fluctuations in tankship rates.[15] Recent events had demonstrated that such a settlement was the only method of preventing the "leapfrogging" of demands for revenue increases: the original Libyan settlement had triggered the Caracas demands, and Libyan officials had already made clear that any revenue increases obtained by the

---

[13] For details, see statement of Thomas E. Kauper, 5 July 1974, in *MNC Hearings* 9: 46-49.

[14] Letter to Richard W. McLaren, U. S. Department of Justice, from John J. McCloy, 19 October 1971, in *MNC Hearings* 6: 231-46.

[15] For full text of the letter, see *ibid.*, Exhibit 1, p. 60.

Persian Gulf producers would also be demanded by Libya.[16] Conducting negotiations on a global basis was the only method of ironing out the differentials in per-barrel taxes which gave rise to new demands.[17]

Libyan officials soon learned of the oil-sharing agreement, and met with representatives of the independent companies in January 1971. The Libyan oil minister made a number of threats to various independent companies which culminated in an ultimatum to Hunt and Occidental to dissociate themselves from the group of companies by 24 January 1971 or suffer "appropriate action." Although Hunt and Occidental expected that new production ceilings would be imposed, they refused to leave the LPG or to accept higher per-barrel taxes. Libya took no action. Apparently, the oil-sharing agreement had made the independents considerably less vulnerable to Libyan tactics.

2. U. S. Intervention and the Structure of Negotiations. Undersecretary of State Irwin began a round of visits on 16 January 1971 with the Shah of Iran, King Faisal of Saudi Arabia, and the Emir of Kuwait. The purpose of the mission, according to Irwin, was to seek assurances from the Persian Gulf producers to continue "to supply oil at reasonable prices to the Free World."[18]

It is possible that a strong stance by the United States against price increases would have been a sufficient counterweight to OPEC's growing power. However, the Department of State lacked a coordinated approach to the pricing issue. It was preoccupied by more limited objectives, such as maintaining friendly ties with the Shah, or ensuring that the Saudi government had sufficient revenues to maintain a strong role in the Persian Gulf.[19] It saw the oil supply issue as only one aspect of the general political situation in the Middle East.

[16] Statement by G. H. M. Schuler, *ibid.*, p. 7.

[17] Again, parallels to industrial union bargaining are striking. Wage settlements are frequently higher when different unions negotiate on contracts at different times because wage gains in previous bargaining become the starting point for ensuing negotiations. Since the producers sought to maximize the amount of the settlements, geographically separate negotiations which would be concluded at different times very much suited their objectives.

[18] Testimony of John J. Irwin II, 31 January 1974, in *MNC Hearings* 5: 147-49.

[19] See British Ambassador's Report to London on MacArthur Briefing of 19 January 1971, *ibid.* 6: 65, Exhibit 4.

THE SHIFTING BALANCE OF MARKET POWER

Irwin and Douglas MacArthur II, the American ambassador to Iran, met with the Shah and with Minister of Finance Dr. Jahangir Amuzegar on 18-19 January. After conversations with the Shah and his finance minister, Ambassador MacArthur strongly recommended to Washington that the companies be advised to begin negotiations with the Persian Gulf group immediately and to hold parallel negotiations with Libya. MacArthur's basic concern seems to have been to maintain good relations with the Shah, since MacArthur reported that he was "deeply suspicious and afraid that the companies intended to play OPEC members against one another."[20] Ironically, the companies' ability "to play OPEC members against one another" was the whole basis of the companies' negotiating strength. After MacArthur's recommendation, Secretary of State William P. Rogers communicated the support of the U. S. government for separate negotiations in the Persian Gulf, which were further supported by the British ambassador.[21]

On 19 January 1971 the co-chairmen of the company negotiating team—Lord Strathalmond and G. T. Piercy—met with the Iranian Minister of Finance Amuzegar, Saudi Oil Minister Ahmed Zaki Yamani, and Iraqi Oil Minister Sa'dun Hamadi. The two company negotiators concluded that the Iranians were convinced that the American government was willing to abandon its demands for OPEC-wide negotiations.[22] Amuzegar escalated the level of threats during the next ten days, and repeated demands that negotiations begin immediately with the Persian Gulf producers. With U. S. support withdrawn, the company representatives in London were faced with a difficult decision. If the companies continued to insist upon global negotiations, the dialogue with the Persian Gulf producers might well end. The companies had been informed that they would receive little support from the consuming countries if the cause of a breakdown in negotiations were the continued insistence by the companies on OPEC-wide negotiations.[23] At the 19 January meeting, Amuzegar had given the companies a 48-hour ultimatum to agree to "Persian

[20] Inward Message (from Teheran Negotiating Team to LPG), in *MNC Hearings* 6: 64, Exhibit 4.

[21] British Ambassador's Report to London on MacArthur Briefing on 19 January 1971, *ibid.*, p. 66, Exhibit 4.

[22] Message No. 2 from Teheran Passed through West End Friends (British Embassy), 19 January 1971, *ibid.*, p. 69.

[23] Standard Oil Company of California, Submission for the Record in Extension of Oral Testimony Taken on the Afternoon of March 28, 1974, *ibid.* 8: 666.

Gulf only" negotiations. If the companies refused, the Iranian minister of finance threatened to hold an OPEC conference on 25 January, at which "Venezuelan terms" would be legislated by the entire OPEC membership. The LPG negotiators, meeting in Teheran on 21 January, were confronted with the alternatives of discussing substantive issues relating only to the Persian Gulf countries or of rupturing the negotiations. Piercy stated that he did not wish to end the negotiations on a matter of procedure, and began discussing substantive issues.[24] It was agreed that negotiations would be finished by 1 February.

The Iranian threats were taken seriously. It was widely recognized that the Libyan settlement in October resulted in per-barrel increases greater than any market advantages of Libyan crude. It was conceivable that OPEC might unilaterally increase the tax rate to 60 percent without the necessity of regulating production. Strathalmond and Piercy summarized:

> We are far from home particularly in regard to a Gulf settlement not being dependent on a settlement elsewhere. We are sure this will not, repeat not, be agreed to here and if we push for global settlement this will only lead to a demand for Venezuelan terms everywhere.[25]

Believing that industry-wide cooperation was the only basis of strength, the LPG, however, refused to abandon the joint negotiating stance.

At the 19 January meeting, Amuzegar stated that the oil ministers of the Persian Gulf countries would write a letter to the companies specifying the conditions for subsequent negotiations. The LPG instructed their representatives in Teheran to do everything possible to prevent such a letter, and further instructed that if such a letter were delivered, the LPG representatives should reply to it by restating the companies' position that a Persian Gulf settlement would not be acceptable if negotiated independently of other producing areas.[26]

At this point, Sir Denis Wright, the British ambassador to Iran, and American ambassador MacArthur directly intervened and con-

---

[24] Inward Message VR0005 (from Teheran Team to LPG), 21 January 1971, in *MNC Hearings* 6: 78, Exhibit 16.

[25] *Ibid.*

[26] Outward Message JA0003 (from LPG to Teheran Team), 22 January 1971, *ibid.*, p. 79, Exhibit 20.

vinced Strathalmond to ignore his instructions. Strathalmond was told by the two ambassadors that there were strong suspicions that the companies were only stalling until a new Libyan settlement could be reached and that rejection of the Persian Gulf producers' conditions would "only finish everything here and now."[27]

The intervention of the British and American officials was clearly decisive in the final decision of the companies to abandon their joint negotiating stance. The American officials initially made clear to the Persian Gulf producers that they were willing to drop their support of OPEC-wide negotiations, and this induced the producing governments to take an even stronger stand against the companies' position. When the LPG decided to reject the ultimatum for separate negotiations from the Persian Gulf producers, the British and American ambassadors prevented the LPG's representative from carrying out his instructions.

The companies were now forced to adopt separate—but what they hoped would be simultaneous—negotiations, and the original Teheran team was split into a Persian Gulf unit (headed by Strathalmond) and a Libyan unit (headed by Piercy). The strategy of the LPG now called for the two units to serve as sectors of one common front. However, neither unit contained a representative of a company which did not operate in the area where negotiations were to take place, which made the global approach a fiction.

The companies still thought that in the Persian Gulf negotiations it might be possible to provide a "hinge" for a stable Libyan settlement by establishing the level of taxes on Saudi and Iraqi crudes that were exported from Mediterranean terminals, but this strategy failed. The Tripoli negotiating team was received by Oil Minister Izz Din Mabruk on 28 January, but the minister refused to consider the industry proposal. By 30 January, the Tripoli team determined that the RCC was merely waiting for the settlement in Teheran, and the team flew back to London. The attempt to fix taxes on the Mediterranean exports of Persian Gulf crudes to provide a benchmark for Libyan crudes also failed. Strathalmond was informed by Iranian officials that the Persian Gulf producers would not negotiate on Mediterranean exports.

In sum, the joint negotiating stance was abandoned largely because of the pressure of the United States and Great Britain. Thereafter, the French, Belgian, and Japanese companies with large

[27] Inward Message VR0006 (from Teheran Team to LPG), 22 January 1971, *ibid.*, p. 79, Exhibit 21.

Persian Gulf operations lost interest in the Libyan settlement. The LPG, therefore, lost the potential support of a number of consuming-country governments in their negotiations with Libya. Once the companies abandoned the strategy of joint negotiations, none of their other strategies succeeded because the companies were forced to rely upon the cooperation of the producing governments. For example, promises by the Persian Gulf states to include crudes exported from the Mediterranean in the Persian Gulf negotiations were subsequently ignored. One of the most important consequences of the success of the Persian Gulf producers in preventing an industrywide effort at solidarity was psychological: the apparent inability of the companies to maintain a joint front under threats of punitive legislation, nationalization, and OPEC-wide limitations on production made it probable that such threats would be used again by the producers.

B. FINAL SETTLEMENT WITH THE GULF PRODUCERS

Once the industry's initial negotiating stance collapsed, a settlement which was highly favorable to the Persian Gulf producers was inevitable. Only the final terms of settlement and their impact on subsequent Libyan demands remained to be determined. Table 2 summarizes: (a) the LPG negotiating team's original offer of 28 January 1971, (b) the original demands of the Persian Gulf producers' committee, and (c) the terms of the 14 February Teheran settlement. Since the agreement was to include an escalation clause, Table 2 shows anticipated revenue increases over a five-year period.

After the LPG made its initial offer on 28 January, there were two days of negotiations in Teheran punctuated by additional threats of expropriation and of an oil embargo. The companies were seeking to avoid any precedents which might prove damaging in the forthcoming Libyan negotiations, and to determine a minimum per-barrel increase that would avoid the unilateral imposition of new taxes. It soon became clear that the original offer of the companies was below this minimum, and a new offer to increase posted price by 20 cents/bbl. was made on 31 January 1971.

On 1-2 February, diplomatic representatives of the United States and the other major consuming countries met with the Shah and with other producing-government officials in Teheran in an attempt to moderate the demands which the Shah was expected to make at the OPEC conference on 3 February. Considering the threats made by OPEC at that conference, one cannot characterize the intervention of the consumers as successful. The U. S. Department of

THE SHIFTING BALANCE OF MARKET POWER

Table 2

BASIC TERMS OF ORIGINAL LPG OFFER, GULF PRODUCERS' DEMANDS, AND FINAL TEHERAN AGREEMENT FOR 1971-1975

(In cents per barrel)

Terms[a]	Year				
	1971	1972	1973	1974	1975
Original LPG Offer					
Posted price increase	15	15	15	15	15
Additional allowance (incl. inflation)	0	1	3	5	7
Total posted price increase	15	16	18	20	22
Increased government take	9.1	9.7	10.9	12.1	13.4
Gulf Producers' Demands					
Posted price increase	54	54	54	54	54
Additional allowance (incl. inflation at 4.5 percent)	20	33	48	63	78
Total posted price increase	74	87	102	117	132
Increased government take	45.0	52.8	61.8	70.8	80.0
14 February Teheran Agreement					
Posted price increase	35	35	35	35	35
Additional allowance (incl. inflation at 2.5 percent)	16	19	28	38	48
Total posted price increase	51	54	63	73	83
Increased government take	30.0	33.0	38.0	44.0	50.0

Source: Letter to Richard W. McLaren, U.S. Department of Justice, from John J. McCloy, 23 July 1971, in *MNC Hearings* 6: 240.

[a] Estimated financial costs for average Persian Gulf oil.

State pressured the companies throughout the negotiations to obtain assurances from the Persian Gulf producers that oil would continue to flow "at fair prices." For example, McCloy informed the companies that the "higher echelons" of the State Department were hoping "that the companies would get across the idea that they were prepared to explore all avenues to get the proper assurances."[28]

One representative of the LPG reported that three possible strategies were being considered in London in February before negotiations were temporarily disrupted: (a) acceptance of the producers' demands and of assurances that the companies did not deem meaningful; (b) refusal to offer any more in taxes per-barrel and rupture of the negotiations; and (c) achievement of a stable settlement by offering to pay more taxes per-barrel, conditioned upon inclusion of Mediterranean crudes in the negotiations.[29] The third strategy was another attempt to maintain the "hinge"—the posting of Persian Gulf crudes exported from the Mediterranean—that would provide a basis for negotiations with Libya, albeit at higher per-barrel taxes. Instead of making offers of higher per-barrel revenues in order to obtain a benchmark that might prove useful in moderating Libyan demands, the companies paid more in exchange for promises by the Persian Gulf producers to maintain price stability after the Libyan settlement and to refrain from imposing limitations on oil production. The companies allowed themselves to be persuaded by the governments of the consuming countries to settle for assurances, even though experience had taught them that such assurances were of little value. The assurances of the Persian Gulf producers were not meaningful because it was not in their self-interest to adhere to them, and they could not be enforced. Iraq and Saudi Arabia, for example, had an interest in a higher Libyan settlement because such a settlement would increase their revenues on exports shipped from Mediterranean terminals.

The crucial factor determining the final terms of settlement in Teheran was the inability of the consuming countries to organize any resistance to the demands of the producing countries. On 20 January 1971, U. S. officials met in Paris with representatives of the other OECD nations (with company representatives present as observers) to discuss the Teheran negotiations.[30] The U. S. Department of

[28]McCloy to Lindemuth Message Transmitted to Moses, 1 January 1971, in *MNC Hearings* 6: 101, Exhibit 44.

[29]Statement of G. H. M. Schuler, *ibid.*, p. 35.

[30]For a summary of the press reports on this meeting, see Adelman, *World Petroleum Market*, pp. 254-55.

State also established an interagency task force in early January to consult with European governments.[31] During the meeting and the consultations, no serious attention was paid to the development of plans for oil-sharing, or for increasing stockpiles, or for managing stockpiles in the event of an OPEC-wide shutdown. George Schuler, the LPG representative of the Bunker Hunt Company, summarized the consequences of this inaction as follows:

> [T]he producing states, once they recognized that they were not going to be restrained by the joint action of the industry, and that the governments were not going to support the industry, then the only thing to hold them back was their own self-restraint.[32]

The February deadline for the termination of negotiations set by the Persian Gulf producers passed; no agreement was reached on assurances against leapfrogging or on financial terms, and the negotiations were terminated. An extraordinary conference of OPEC was held in Teheran on 3-4 February. The organization adopted a resolution calling for the introduction on 15 February 1971 of "the necessary legal and/or legislative measures" if previous demands were not met.[33] The resolution further stated that, in the event of noncompliance by the companies, all crude oil and products shipments would be embargoed.[34]

OPEC's deadline put additional pressure on the companies to settle. The final agreement, signed 14 February 1971, was to be effective until 31 December 1975. The agreement specifically prevented the renewal of demands after any Libyan settlement and contained assurances that there would be no action in support of Libyan demands.

C. THE SETTLEMENT WITH THE MEDITERRANEAN EXPORTERS

While no general agreement was reached in Libya, the settlement between the government and each of the operating companies was basically similar. Most agreements were signed on 20 March 1971 and generally provided for: (a) an increase in the posted price of 40°

---

[31] Testimony of John W. Irwin II, in *MNC Hearings* 5: 147.

[32] Testimony of G. H. M. Schuler, 31 January 1974, *ibid.*, p. 123. Piercy agrees with this assessment, although he stresses that economic factors played a role as well (Testimony of George Piercy, 1 February 1974, *ibid.*, pp. 243-44).

[33] *OPEC Resolution* XXII.131.

[34] *Ibid.*

API crude, escalating from 1973 through 1975, including a 12 cents/bbl. Suez canal allowance and a 13 cents/bbl. freight premium; (b) a uniform tax rate of 55 percent; and (c) certain supplemental payments for reinvestment in Libya. The companies also agreed to continue exploration activities. The net effect of the settlement—exclusive of the escalation clause or of company exploration commitments—was to raise government take by 65 cents/bbl.

The strategy[35] of the RCC was basically similar to that employed in the negotiations of 1970. The Libyans initially refused to negotiate with a group of companies in the hope of again exploiting the weaknesses of the independents. The operating companies, however, continued to tender identical offers, and the Libyans were forced to accept negotiations which were essentially collective, although the Libyans demanded that at least one of the two company representatives in the final negotiations be a representative of an independent company, because the RCC believed that the independents were more vulnerable.

Threats of nationalization and of production cuts were issued at regular intervals by the RCC. The threat of nationalization was made more credible by close cooperation between the Libyan and Algerian governments. In February 1971, Algeria assumed control over 51 percent of French producing interests (accounting for some 70 percent of Algerian production). The Algerians had granted compensation, but withheld most of it as an enforced payment for higher taxes. Libyan threats of a production shutdown were also supported by a joint communique signed by Iraq and Algeria on 25 February 1971 which stated that if the companies refused to accept the minimum Libyan demands, there would be a curtailment of oil production by all major Mediterranean producers.[36]

The objectives of the majors and of the independents in the second round of Libyan negotiations were largely compatible. The majors desired to minimize per-barrel increases because of the possibility that higher increases would lead to a leapfrogging of demands

---

[35]While the protracted negotiations in Tripoli provided a number of typical examples of the tactics used in distributive bargaining, our primary concern here is the strategic behavior of the actors. Libyan negotiators particularly excelled in the use of what Walton and McKersie call "commitment tactics" (Richard E. Walton and Robert B. McKersie, *A Behavioral Theory of Labor Negotiations: An Analysis of a Social Interaction System* [New York: McGraw Hill, 1965], pp. 82-121).

[36]Joint Communique—February 25, 1971, in *MNC Hearings* 6: 188, Exhibit 91.

into the Persian Gulf. The majors, however, were more prone than the independents to grant higher increases as long as the increases were part of the fluctuating freight element of Libyan per-barrel taxes. On the other hand, the independents—who continued to depend exclusively upon Libyan oil—desired to limit all increases, whether permanent or fluctuating. There was considerable conflict over this issue within the LPG when it formulated the terms of its offer, but the companies compromised. The independents had no alternative but compromise. If they withdrew from the LPG and the "safety net" of the oil-sharing agreement, they would be completely vulnerable to Libyan demands. The final deliberations of the companies on 12 March 1971 over settlement with the Libyan government were summarized on a blackboard at Britannic House in London as follows:

1. Pro Settlement:
    (a) Defers nationalization and makes it less tempestuous
    (b) Helps to maintain the industry as the middleman
    (c) Keeps the oil flowing

2. Con Settlement:
    (a) Puts strains on Teheran
    (b) Makes Libyan oil uncompetitive
    (c) Whets appetite in Libya and elsewhere
    (d) Possibly difficult to convince Europe of its necessity
    (e) Strengthens appeals of Libyan-type approach[37]

Points (a) through (c) under "Con Settlement" indicate that industry executives had a good idea of subsequent developments. Shortly after the Tripoli agreement, Minister of Finance Amuzegar told representatives of the Iranian consortium that the Shah was "incensed" that the industry had given Libya more than the Persian Gulf countries, and had threatened to lend the support of Iran to "radical" OPEC members if the consortium would not make additional tax concessions to Iran.[38] The basic terms of the Libyan settlement were applied to Saudi Arabian and Iraqi Mediterranean exports in June 1971; the increase in government revenues was approximately 80 cents/bbl.

---

[37]Statement of G. H. M. Schuler, *ibid.*, p. 40.

[38]Letter to Richard W. McLaren, U. S. Department of Justice, from John J. McCloy, *ibid.*, p. 231.

D. CONCLUSION

The increase in the market price of crude during 1970-71 reflected successful bargaining by the producing governments on per-barrel taxes. The explanation of that success lies in the objectives and the strategies of the major actors. The producing governments shared the objective of increasing their revenues from oil production, and the strategies employed by the "radical" government in Libya and the "conservative" governments in the Persian Gulf were basically similar. However, one must not overstate the degree to which producers were willing to cooperate on the imposition of sanctions: a joint threat is not an embargo. Both Saudi Arabia and Iran were busy pressing the operating consortiums to increase the volume of output during 1970-71, and if production cutbacks had taken place, there would have been considerable temptation in other countries to use them as an opportunity to increase market shares and revenues. In the Teheran/Tripoli negotiations, the strategies adopted by the oil-exporting countries produced an increase in their collective power over the pricing of crude oil. There remained, however, a considerable potential for the exercise of monopoly power, but this potential could not be exploited until the imposition of more extensive limitations on production.

Despite recent claims that the companies combined with the producers to raise prices, the evidence supports the view that a basic objective of the operating companies in the Teheran/Tripoli negotiations was to minimize tax increases.[39] Increases in tax payments to producing governments in 1970 only added to excess, and therefore unusable, foreign tax credits of American companies, since the tax rate on "profits" from production then exceeded the U. S. corporate tax rate. As a Socal official explained:

> [E]ach additional dollar paid to a foreign country reduces our profit margin unless we can recover it by increasing the market price, which represents our ultimate source of revenue.[40]

The companies later managed to recover the additional taxes and increase margins by increasing product prices, but this profit recovery was determined by market conditions and by tacit industry price

---

[39] After extensive hearings, the Senate Subcommittee on Multinational Corporations reached the same conclusion (*MNC Hearings* 7: 279).

[40] Standard Oil Company of California, Submission for the Record, in *ibid*. 8: 667.

collusion.[41] During the 1970-71 negotiations, however, the impact that higher taxes would have on company margins was still uncertain. The commitment of the companies to minimize posted price increases was, of course, not a total commitment. A proportion of company assets were in the producing countries with which they negotiated. As we have noted, the companies attempted to allow only the smallest increases deemed necessary to deter producing governments from nationalization or expropriation of their assets. The most fundamental objective of the major oil companies, therefore, was probably to avoid a disorderly market which would threaten earnings and market shares. This objective accounts for the insistence of the majors on joint negotiations, and it also accounts for the willingness of the majors to grant revenue increases to Libya as long as tax-paid costs between the Persian Gulf and the Mediterranean remained harmonized.

The companies were more prepared for a confrontation with producing governments over revenues than were the governments that represented the ultimate consumers of oil. Negotiations which began as a commercial dispute rapidly became politicized, and the U. S. government played a key role in the outcome. The U. S. Department of State saw the oil negotiations as only part of its policy in the Middle East. There was concern among U. S. officials that Iran and Saudi Arabia might need additional revenues to play a strong anti-Communist role in the Persian Gulf. The insistence of the Department of State that the companies obtain "assurances"— even at the cost of higher per-barrel revenues—is consistent with a general "diplomatic orientation" to the oil problem. The Department of State followed the principle of making it financially advantageous for another government not to take action, rather than to threaten or to use force. What industry executives realized, because of their long experience, was that the producers' demands were governed by "commercial" rather than "diplomatic" logic.

The success of the producing governments encouraged their further collective efforts and demands. Also, the higher revenues from the settlement made it possible for the producing governments to impose limitations on production for longer and longer periods. Although the Department of State insisted upon viewing the oil negotiations in a wider political context, the basic objective of even the "radical" regime in Libya was an increase in per-barrel revenue. The short-term objectives of the Department of State (such as

[41] See p. 75 below.

maintenance of good relations with the Shah) resulted in a long-term shift to the producing states in the balance of power.

It is sometimes claimed that American officials were quite willing to accept higher oil prices.[42] If oil prices increased, higher energy costs would be imposed largely upon European and Japanese industry, which would make American goods more competitive. While there is insufficient evidence to determine whether this factor entered into the deliberations of the Department of State, subsequent events showed that the inaction of consuming governments limited their power over oil pricing.

---

[42] Suggested in "The Phony Oil Crisis: A Survey," *The Economist*, 7 July 1973, p. 19. Similar charges were made by a CFP official at the time of the negotiations, and were reported by McCloy to the Department of State (see Memorandum for Files re: Oil Companies' Negotiations with OPEC, 3 February 1971, in *MNC Hearings* 6: 300).

Chapter V

THE BREAKDOWN OF COMPANY-GOVERNMENT BARGAINING, 1972-1973

After the Teheran and Tripoli agreements, the balance of power in the market for Middle Eastern crudes continued to shift to the producing countries. There were three major developments in international bargaining over oil prices and production levels during 1972-73. First, despite the appearance of continued bilateral negotiations between companies and governments, unilateral decisions of the producing countries became the most important determinants of price. Second, bargaining over price became firmly linked with the wider political conflict between the United States and many Arab states. Third, cooperation among OPEC members played a more important role in price determination. As a result of these developments, government revenue per-barrel increased throughout the period. (Table 3 summarizes the fiscal changes which took place in Saudi Arabia, the largest Middle Eastern producer, from 1970 to 1973.)

A. SUPPLY, DEMAND, AND PRODUCTIVE CAPACITY

There were a number of important changes in the supply and demand characteristics for Middle Eastern crudes in the 1972-73 period. The demand for energy is highly income elastic, and in the early 1970's most of the developed world was in a period of high employment and economic growth. Growth rates in real GNP for the developed world increased in 1970-73 as shown in Table 4.

Most of the accompanying rise in energy consumption was met by an increase in the production of oil from Middle Eastern sources. The demand for Middle Eastern oil was further increased by the emergence of the United States as a substantial importer.[1] Crude oil imports into the United States rose by 26 percent in 1971,

[1] See Edward W. Erickson and Leonard Waverman, "Introduction," in *The Energy Question: An International Failure of Policy*, eds. E.W. Erickson and L. Waverman (2 vols.; Toronto: University of Toronto Press, 1974), 1: xix.

Table 3

GOVERNMENT REVENUE PER-BARREL FOR SAUDI ARABIAN 34° API MARKER CRUDE: 1970-1973

Date	Dollars per Barrel[a]	Explanation of Change
14 November 1970	$0.98	
15 February 1971	1.27	An increase in posted price and a reduction in discounts resulting from the Teheran/Tripoli agreements, which contained an inflation index.
20 January 1972	1.45	An increase in posted price resulting from the Geneva I agreement, which was designed to compensate for the devaluation of the U. S. dollar.
1 January 1973	1.62	An escalation of posted price as specified in the Teheran/Tripoli agreements, and the effect of the General Agreement on Participation.
1 April 1973	1.73	An escalation of posted price resulting from the Geneva I agreement.
1 June 1973	1.82	An increase in posted price resulting from the Geneva II agreement, which was designed to compensate for further devaluation of the U. S. dollar.

Source: U. S. Department of the Interior, Office of Oil and Gas, *Worldwide Crude Oil Prices* (Washington: Department of the Interior, 1973), Table 5 and Appendix A.

[a] All calculations assume a producing cost of 10 cents/bbl.

Table 4

REAL GNP INCREASES FOR DEVELOPED WORLD: 1970-1973

Year	Percent Increase
1970-1971	3.6%
1971-1972	5.5
1972-1973	6.3

Source: Darmstadter and Landsberg, "The Economic Background," p. 17.

and 17.5 percent of such imports came from the Middle East. During 1972, imports of oil into the United States rose by 32 percent.[2]

The expansion of oil imports into the United States was a reflection of a number of U.S. energy policy decisions and setbacks and of a heightened concern with environmental protection objectives. For example: (a) the price of natural gas was set so low that natural gas shortages developed, further increasing oil demand; (b) the government denied oil companies access to producing areas offshore; (c) there were delays in the construction of the Alaska pipeline; (d) nuclear power developed more slowly than expected; and (e) the production of coal was allowed to decline.[3]

The key factor influencing the bargaining position of the producing governments was the disappearance of spare capacity for oil production outside of the Middle East. The trend in excess capacity in the non-Communist world during 1960-73 is shown in Figure 3. Figure 3 indicates that by 1970 almost all of this excess producing capacity was located in countries other than the United States; in the United States, excess capacity had virtually disappeared by 1972. (The crosshatch area of the figure shows spare capacity that was unusable because of production restrictions imposed by the governments of Venezuela, Kuwait, and Libya prior to October 1973. Such production limitations further reduced the amount of excess capacity available to the international oil companies.)

Although there were problems in expanding the production capacity of even the richest Middle Eastern reserves,[4] Middle Eastern production increased rapidly. During 1970-71 production increased by 17 percent; during 1971-72 production increased by 11 percent and Saudi Arabian output rose by 28 percent.[5] In the first seven months of 1973, Middle Eastern output rose by 23 percent.

B. PRICES AND REVENUES

Between the spring of 1970 and the summer of 1973, market prices of Middle Eastern crudes more than doubled when measured in dollars and nearly doubled when measured in terms of other

[2]Darmstadter and Landsberg, "The Economic Background," p. 22.

[3]Herman Kahn, *Oil Prices and Energy in General* [Hudson Institute Paper HI-2063/3-P] (Croton-on-Hudson, N.Y.: Hudson Institute, August 1974), p. 4.

[4]Hartshorn, "Oil Diplomacy," p. 283.

[5]*BP Statistical Review*, various issues, 1970-73.

Figure 3
ESTIMATED NON-COMMUNIST WORLD SPARE CRUDE OIL PRODUCING CAPACITY: 1960-1973

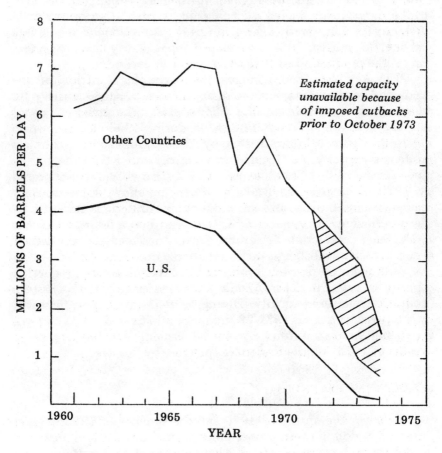

Source: Adapted from Exxon's estimates as reported in Statement of G. T. Piercy, in *MNC Hearings* 7: 335.

major currencies. Only half of this increase was due to higher producing-government taxes.[6] Since the real costs of production did not increase substantially, the higher prices for crude and petroleum products resulted in higher oil company profits. It has been estimated that major oil company profits increased from 30 cents/bbl. in 1971 to 90 cents/bbl. in the spring of 1973.[7]

Negotiations between companies and producing governments regarding participation in oil concessions began in early 1972. The majors were not certain about the quantity of oil that would have to be allocated to the national oil companies of the producing countries or about the effect on tax-paid costs. Therefore, they reduced their sales of crude to third parties, and new buyers were forced into the market. Purchasers of crude, such as the independent refiners, discovered that they could no longer obtain long-term supply contracts, and bought increasing amounts of crude in the spot market at higher prices; these developments resulted in competitive overbidding and increased market prices.

Independent oil companies also submitted competitive bids for the limited amount of participation crude which became available in the spring of 1972. By May 1972, Saudi Arabia had sold all of its participation crude available for 1973 and part of that to become available in 1974 and 1975. Although these sales involved relatively small amounts of oil, the national oil companies obtained "record prices."[8] This competitive bidding was further accelerated by the U.S. government's removal of all quantitative restrictions on oil imports as of 1 May 1973. Thereafter, American inland refiners joined the Europeans and Japanese in attempting to obtain participation crude. Even American utility companies sent missions to the Middle East to attempt to negotiate supply contracts. By August 1973, market prices for Mediterranean crude had risen well above the posted prices.[9]

Finally, increasing demand for Middle Eastern oil had a significant impact on the tanker market. Demand for tankers rose

---

[6] Edith Penrose, "Development of Crisis," p. 47.

[7] Testimony of Dillard P. Spriggs, 28 January 1974, in *MNC Hearings* 4: 61. Spriggs's figure for 1971 is identical to the estimate made by Parra, Ramos, and Parra, *International Crude Oil and Product Prices*, 15 October 1971, p. v, and his figure for 1973 is similar to the estimate made by *PIW*, 20 August 1973, p. 34.

[8] Penrose, "Development of Crisis," p. 48.

[9] *Ibid.*, p. 49.

substantially in late 1972, and expanding U. S. import demand sent spot rates up to new record levels in the first half of 1973. (For example, rates rose from 74 cents/bbl. in April 1972 to $4.09/bbl. in August 1973 for a single Persian Gulf-Rotterdam voyage.)[10]

These changes in market conditions had important implications for company-producing government negotiations on price:

> First, the substantially increased profit margins of the major oil companies from late 1971 to mid-1973 made new demands for an increase in per-barrel taxes more likely. Thus, in the summer of 1973, OPEC members renewed demands for higher taxes. They argued that the division of revenues between producing countries and companies at the time of the Teheran/Tripoli agreements was 80 percent to 20 percent in favor of the producers, but that higher market prices had changed this division to 60 percent to 40 percent.[11]
>
> Second, because sales of participation crude provided an immediate indicator of both higher market prices and higher oil company margins and because the producing governments had little difficulty in selling their share of crude, the belief of these governments that they could assume much greater control over the exploitation of their resources was strengthened.
>
> Third, because of the disappearance outside of the Middle East of spare productive capacity, the negotiating position of the major producing countries was decisively strengthened. Moreover, for the first time the United States was dependent upon Middle East sources to meet an increasing portion of domestic demand.
>
> Finally, in a period of high freight rates, it was impossible for the companies to resist the demands of the Mediterranean producers by increasing Persian Gulf production. This helps to explain the success of Libya in marketing nationalized oil in mid-1973, and the ability of the Algerian government to raise posted prices unilaterally to $5/bbl. in September 1973.

C. THE EVOLUTION OF COMPANY OBJECTIVES

The Teheran/Tripoli agreements produced an important change in the objectives of the companies. In 1971 the companies found that they could pass on negotiated tax increases to the consumer

---

[10] *PE*, various issues, 1972-73.

[11] Prepared statement of D. P. Spriggs, in *MNC Hearings* 4: 77.

THE BREAKDOWN OF COMPANY-GOVERNMENT BARGAINING

and at the same time increase their margins. Adelman has suggested that the uniform tax settlement provided a form of natural price leadership to which the companies could respond without collusion.[12] The companies realized that if they acted in concert after a new round of tax increases, they could increase their margins by raising product prices—at least temporarily—until such prices were eroded by competition.

As a result, company executives and even producing government officials reinterpreted the role of the international company. For example, OPEC Secretary-General Adnan Musahim Pachachi said in 1971 that he saw no basic conflict between the companies and the producing governments.[13] Also, Sir Eric Drake, chairman of BP, reported to the company's shareholders in 1970 that the "industry has become a tax collecting agency, and incidentally how poorly rewarded it is for doing so, bearing in mind the enormous risks it has to take."[14] The major oil companies were resigned to the fact that most of the higher profits that resulted from the tighter supply situation from late 1972 onward would primarily benefit the producing governments.[15] Since the consumers were willing to pay higher prices, the companies increasingly identified their own interests with those of the producing governments.[16] One might speculate that the warnings made by company executives about an impending "energy crisis" beginning in 1972

[12] Adelman, *World Petroleum Market*, p. 252.

[13] *New York Times*, 11 February 1971.

[14] *BP Annual Report and Accounts for 1970* (London: BP, 1970), p. 5.

[15] Peter R. Odell, *The Western European Energy Economy: Challenges and Opportunities* [Stamp Memorial Lecture 1975] (London: Athlone Press, 1975), pp. 14-15, and P. R. Odell, *Oil and World Power: Background to the Oil Crisis* (3d ed; Harmondsworth: Penguin, 1974), pp. 200-201.

[16] A Senate committee staff report concluded early in 1973 that the petroleum industry could not be expected to adopt policies which would restrain tax increases: "As long as the tax increases (or conditions for 'participation') by different exporting nations are enforced equally against the various companies, they can (and must) pass the major burden of increased costs on to consumers" (U. S. Congress, Senate, Committee on Interior and Insular Affairs, *Towards a Rational Policy for Oil and Gas Imports*, p. 10). A director of Shell made the same point, although in a different context: "What fact is important to the oil company is the *margin* between costs and proceeds, not the absolute level of price, except insofar as the level of price ultimately affects volume and therefore margins" (Geoffrey Chandler, "The Role of the International Oil Companies," address at the Second International Colloquium on the Petroleum Economy, Laval University, Quebec, 3 October 1975; emphasis in original).

were designed to prepare citizens in the consuming countries for the higher oil prices that the companies anticipated would follow the increasing shift in power to producing governments.[17] The companies expected that under tighter market conditions their higher margins would result in the imposition of higher taxes by the producing governments. Thus when the companies realized higher margins in late 1972, they manipulated transfer prices to attribute a higher profitability to "downstream" operations of refining and marketing as a defensive move.[18]

D. CURRENCY FLUCTUATIONS AND TAX INCREASES

The outcome of negotiations on higher posted prices during 1971-73 reflected OPEC's increasing collective bargaining power, and the continuing instability of international money markets. In the summer of 1971, the depreciation in the external value of the U. S. dollar raised new demands for higher taxes because the Teheran/Tripoli agreements fixed the tax reference price in terms of dollars. Libya acted first. On 16 August 1971, the Bank of Libya revalued the non-convertible Libyan dinar, which added 3.5 percent to the dollar cost of taxes and royalties. The OPEC conference of 22 September 1971 demanded action to correct the effect of the depreciation of the dollar that followed the floating of the dollar beginning on 15 August 1971. However, the Libyan government was unwilling to await an OPEC-wide resolution of this issue, and on 26 September 1971 the Libyan central bank established a two-tier exchange market which further increased taxes and royalties by 4.8 percent in dollar terms. All companies operating in Libya refused to pay the additional dollar amount of taxes imposed by the 26 September regulations; in turn, Exxon and Occidental were threatened with "administrative attachment" and revocation of their concessions by the Libyan government, which finally attached the funds of Exxon and Occidental on deposit in Libyan banks.

With the exception of Libya, OPEC members reached a settlement in Geneva on 20 January 1972 which resulted in an increase of 8.57 percent in the revenues of producing governments. In their public reports the companies treated the Geneva I settlement

[17] For example, see Robert Brougham, "An Oil Company View," *Middle East Information Series* 23 (May 1973): 61. (Brougham's statement was originally made before the Subcommittee on the Near East of the Committee on Foreign Affairs, U.S. House of Representatives, 2 February 1972.)

[18] Statement of D. P. Spriggs, in *MNC Hearings* 4: 77.

as a "supplement" to the Teheran/Tripoli accords. As a result of the Geneva I settlement, posted prices were to adjust quarterly in accordance with a formula based on fluctuations in exchange rates. Libyan demands were resisted by the companies because payments to Libya higher than those specified by Geneva I would have resulted in further demands by other producers.[19] The companies were able to resist higher Libyan demands because the RCC's bargaining position was temporarily undermined by a drop in tanker rates and by a reduction in the demand for crude oil in mid-1972. After further negotiations, agreement was reached with the Libyan government in May 1973.

The dollar was devalued for a second time in February 1973, and according to the terms of the Geneva I agreement, posted prices were raised by 5.8 percent and producing government revenues increased by slightly more. However, the producing countries claimed that the existing formula did not fully offset the effect of the dollar devaluation, and they demanded further adjustments. The companies announced that they were willing to adopt "minor amendments" to the Geneva I agreement, and offered to raise posted prices by a total of 7.2 percent—an additional 1.4 percent. In June the companies accepted OPEC demands, and government revenue was increased by almost 12 percent over that of January 1973.

By the conclusion of the Geneva II settlement, there was only the appearance of bilateral bargaining over price. Between February and June 1973, Algeria made an effort to have OPEC repudiate the Teheran/Tripoli agreements. Both Libya and Iraq urged the unilateral posting of tax reference prices. Saudi Arabia and Iran, on the other hand, refused to sanction such a unilateral breach of the agreements because repudiating the agreements would violate OPEC pledges of price stability made to the major consuming governments.

In the Geneva I and II negotiations the companies displayed only token resistance to new demands for revenue increases. The companies passed such increases on to consumers in the form of higher product prices. It has been observed that the companies presented the Geneva I and II agreements as supplements to the Teheran/Tripoli agreements, even though both Geneva agreements involved a significant increase in revenues and could have been equally well characterized as breaches of the five-year agreements.[20]

[19] Statement of G.H.M. Schuler, in *MNC Hearings* 6: 50.
[20] Field, "Oil in the Middle East," p. 83.

The increase in the bargaining power of producing governments was not simply a function of tighter market conditions. Despite the trend towards tighter market conditions, there were periods of substantial excess capacity during 1971-73,[21] but such conditions did not result in a return to bilateral bargaining. By the end of 1971, Libyan oil was seriously overtaxed, and production was 500,000 b/d under the production ceiling imposed by the Libyan government. In addition, during the second and third quarters of 1971, the growth rate of oil consumption declined considerably, and the companies had large stocks on hand.[22] However, it was in the interest of both the OPEC members and the major oil companies to limit price competition. An important advantage to the producing countries in "negotiating" with the companies was that such negotiations made tax-paid costs uniform and made competitive tax discounts by producing governments less likely.

E. PARTICIPATION NEGOTIATIONS

Demands for participation in oil concessions by producing governments reflected their long-standing objective of exercising full national control over the exploitation of their resources. However, the important short-term effect of the participation agreements was to increase per-barrel revenues. The objectives of participation by the producing states in oil concessions first emerged during OPEC's June 1968 conference.[23] The OPEC Conference held in Vienna in July 1971 resolved that members should take immediate steps for the effective implementation of the principles of participation.[24] The producing states sought to achieve the following objectives: (a) an increase in government revenues from the production of oil, (b) the assertion of national sovereignty over all aspects of oil production, (c) the use of the technical and logistic capabilities of the international oil companies in order to restrict supply and to raise revenues, and (d) the facilitation of

---

[21] M.A. Adelman, "U. S. Energy Policy," in *No Time to Confuse: A Critique of the Final Report of the Energy Policy Project of the Ford Foundation: "A Time to Choose: America's Energy Future"* (San Francisco: Institute for Contemporary Studies, 1975), p. 34.

[22] Michael Field, "Oil: OPEC and Participation," *The World Today* 28 (January 1972): 11.

[23] *OPEC Resolution* XIV. 90.

[24] *Ibid.* XXIV. 135.

investment in the downstream operations of refining, transport, and marketing.[25] Not all of these objectives could be achieved simultaneously, and the two objectives that were given the greatest priority were (a) and (c)—namely, an increase in government revenue and the continuing use of the capabilities of the oil companies in order to restrict supply.

When the negotiations commenced, there was widespread speculation that the producing governments intended to market their own oil, and that participation would therefore fundamentally transform the character of the international oil companies. Some company executives felt that participation was "calculated to drive up costs, increase the flow of oil to world markets and lower world prices,"[26] but as the negotiations progressed, it became clear that that was exactly what the major producing governments wanted to avoid. For example, OPEC Secretary-General Pachachi stated early in 1972: "We are as concerned with the maintenance of prices as the oil companies."[27]

In the participation negotiations, the producers were represented principally by Minister Yamani of Saudi Arabia, and the companies by the Aramco consortium.[28] Yamani was more than candid when he articulated the objectives of the producing governments in demanding participation in existing oil concessions.[29] According to Yamani, participation was preferable to nationalization because participation would preserve the role of the integrated majors in restricting output and maintaining market prices. On the other hand, nationalization of the producing assets of the companies would "inevitably deprive the majors of any further interest in maintaining crude oil price levels"; if nationalization occurred, the

[25] See U. S. Federal Energy Administration, Office of International Energy Affairs, *The Relationship of Oil Companies and Foreign Governments* (Washington: GPO, June 1975), p. vi.

[26] Field, "OPEC and Participation," p. 10.

[27] Quoted in Maffre, "U. S. Oil Interests in Bargaining with OPEC," p. 812.

[28] "Recent Negotiations: Background Paper Prepared by the Department of State (May 1973)," in U. S. Congress, House, Committee on Foreign Affairs, Subcommittee on Foreign Economic Policy and on the Near East and South Asia, *Oil Negotiations, OPEC and the Stability of Supply*, Hearings, 93d Cong., 1st sess., 1973, p. 239.

[29] Ahmed Zaki Yamani, "Participation Versus Nationalization," in *Continuity and Change in the World Oil Industry*, eds. Zuhayr Mikdashi, Sherril Cleland, and Ian Seymour (Beirut: Middle East Research and Publishing Center, 1970), pp. 211-33. Objectives of OPEC members are summarized in Schurr and Homan, *Middle Eastern Oil and the Western World*, pp. 136-39.

interests of the companies "would be identical with those of the consumers—namely to buy crude oil at the cheapest possible price."[30] Yamani argued that nationalization, therefore, would result in "a dramatic collapse of the price structure," and producing governments would inevitably attempt to sell larger volumes of oil at lower per-barrel taxes.[31] However, participation ensured that the majors would receive a significant share of the monopoly rent from constricted oil supplies, thus making their interests similar to those of the producing governments. At the same time, a basic advantage of participation in existing oil concessions was that participation coincided with the political aspirations of the producing countries to exercise control over their oil resources.[32]

The objectives of the companies in the participation negotiations were to delay execution of an agreement and to secure the best financial terms possible. The basic conflict between the producing governments and the companies, therefore, was not over the higher per-barrel taxes that participation would entail: the conflict was over compensation for the value of the assets transferred and over the maintenance of the tax advantages extended by consuming governments to the companies on "owned" crude.[33] The companies' objectives were of fundamental importance because they directly affected company profitability; therefore, the majors secured permission from the U. S. Department of Justice to meet in London to plan strategy. The negotiations began in January 1972; by the beginning of March they had not produced any results, and OPEC decided to call an extraordinary conference on 11 March 1972 in order to discuss possible sanctions. Shortly before this meeting was held, Aramco notified OPEC of its acceptance of immediate 20 percent participation, but before final agreement was reached in December 1972, King Faisal of Saudi Arabia had made a number of threats.[34]

The final provisions of the General Agreement on Participation were drafted in December 1972 and accepted by Saudi Arabia, Qatar, and the UAE in early 1973. The pricing formula was designed both to enable the majors to retain tax advantages in the consuming

[30]Yamani, "Participation," p. 214.

[31]*Ibid.*, p. 215.

[32]Mana Saeed Al-Otaiba, *OPEC and the Petroleum Industry* (London: Croom Helm, 1975), p. 117.

[33]Stork, *Middle East Oil*, p. 195.

[34]Kubbah, *OPEC*, p. 83.

countries and to make gradual adjustments in supply. The agreement was complex:

> Companies were entitled to 75 percent of total output, or so-called "equity crude." The price of equity crude remained subject to the Teheran agreement. The 25 percent share of production of the producing governments was divided into three categories. The first category was "participation crude," which governments could sell on the open market; participation crude was only 2.5 percent of total output in 1973. The second category was "bridging crude," which producing governments were bound to sell to the companies. Bridging crude was priced at a "quarter way price"—a quarter of the way between tax-paid costs and the posted prices with supplementary per-barrel payments depending upon quality and freight differentials. The third category was "phase-in crude," which the producing governments could require the companies to purchase. Phase-in crude was priced at tax-paid cost with supplementary per-barrel payments depending upon quality and freight. The producing governments' share of production was to increase progressively over the years, resulting by 1982 in the producing governments' ownership of a majority share of total production.[35]

A direct effect of the participation agreement was to increase tax-paid cost by 9 cents/bbl., producing a further violation of the Teheran agreement's pledge of price stability.[36] As noted above, spot sales of participation crude informed producing governments of company margins, and strengthened the determination of the producing governments to set prices unilaterally.[37] Almost all of the oil, however, was still transferred by the same international companies. For example, in 1973 producing governments sold directly only 30 percent of the total amount of participation crude available to them.[38]

The higher tax-paid costs entailed by the participation agreements were extended to other producing areas. Even Iran, which already had 100 percent participation as a result of the nation-

---

[35] For a more detailed summary, see Joachim Hansen, "Neue Ära der internationalen Ölwirtschaft," *Aussenpolitik* 24 (2 Quartal 1973): 201-10.

[36] *PPS*, April 1972, p. 118.

[37] See p. 74.

[38] H. P. Drewry (Shipping Consultants), *Host Government Participation in the Oil Trade*, No. 20 (London: H. P. Drewry, 1974), p. 12.

alization of its oil industry in the 1950's, concluded a twenty-year agreement with the operating companies on 1 March 1973. Under the terms of the agreement oil was to be sold at prices that yielded "essentially the same financial result that would have accrued had Iran entered into a participation agreement similar to the ones originally concluded in the UAE, Qatar, and Saudi Arabia."[39]

F. NATIONALIZATION

There were strong indications even before October 1973 that the oil-producing countries were moving towards the assertion of total control over supply. More extensive government control was imposed not only in the "radical" producing states—Libya, Algeria, and Venezuela—but also in the "conservative" states in the Persian Gulf. More extensive governmental controls were a product of the increasing determination of certain Arab states to use oil exports as a "weapon" to influence U. S. policy in the Middle East and of the growing conviction of oil exporters that their resources were an irreplaceable asset.

1. Nationalization in Libya. The Libyan government on 7 December 1971 promulgated a decree nationalizing BP's producing assets. The timing of the Libyan action is explained by the fact that Gadafi was incensed by the Iranian occupation of three islands in the Persian Gulf. The Iranian occupation was disputed by the governments of two Trucial states which were nominally under British protection. Gadafi claimed that the intervention resulted from British-Iranian collusion.

Representatives of companies operating in Libya met in New York City and agreed to take steps to prevent the marketing of BP's expropriated oil. Bunker Hunt, which held the Libyan concession jointly with BP, agreed to produce no more than its existing share of Libyan oil. The Libyan government did not respond to the protests of the British government, and BP instituted a suit in Italy to recover the first shipment of the expropriated oil for which the new Libyan national company had found a buyer. In the past the institution of such lawsuits by the oil companies had proven to be a deterrent to government nationalization without compensation. Initially, the prospect of additional lawsuits made it very difficult

[39]Mobil Oil Corporation, *Annual Report 1973* (New York: Mobil, 1973), p. 33.

for the Libyan government to market 225,000 b/d of the expropriated oil.

Accordingly, Libyan officials demanded at the end of 1971 that Hunt market BP's expropriated crude. These Libyan demands were supported by threats of nationalization and production restrictions. Encouraged by the companies' agreement to share production, Bunker Hunt continued to refuse to market BP's expropriated oil. However, in the spring of 1972, the Libyan government entered into a number of barter deals with the Soviet Union and other Eastern European countries. These transactions were effective in neutralizing the companies' ability to institute claims in courts in the West to reclaim the expropriated oil. Finally, in February 1973, an Italian court rejected BP's claim of ownership of a shipment of expropriated oil, basing its decision upon a technical ground. Although the judgment was appealed, a large number of independent buyers in the spot market were now willing to risk purchasing the disputed oil.

The Libyan government on 11 June 1973 announced that it had nationalized all of Bunker Hunt's producing assets. The RCC took this step when it realized that there would be little difficulty in marketing the oil. The nationalization of Bunker Hunt, according to Gadafi, was "a warning to the United States to end its recklessness and hostility to the Arab nation."[40] Thus, Libyan nationalization of the assets of BP and of Bunker Hunt was an attempt to use economic coercion to influence U. S. and British foreign policy. It also demonstrated that in tight market conditions the companies had no ability to resist producing-government seizure of their assets.

Participation negotiations between the operating companies and the Libyan government were concluded in mid-August 1973. At that time, Occidental and the Oasis consortium (except for Shell) accepted a 51 percent takeover. The compensation for company assets was lower than that prescribed by the General Agreement of 1972, and buy-back oil was priced at going market prices—i.e., $4.90/bbl. On 1 September 1973, the Libyan government decreed the 51 percent nationalization of the assets of the other operating companies which, unlike Occidental and Oasis, had not accepted Libyan terms. After the 51 percent nationalization of the assets of the majors operating in Libya, the production-sharing agreement among the companies collapsed. Texaco, Socal, Exxon, and BP—which had been supplying Bunker Hunt at tax-paid

---

[40] Statement of G.H.M. Schuler, in *MNC Hearings* 6: 56.

cost—announced they no longer would give assistance.[41] So ended the remaining source of the companies' strength in resisting nationalization.

The companies attempted to enlist the aid of the U. S. Department of State in order to prevent the purchase of the "hot" Libyan oil which had been expropriated from BP and Bunker Hunt. The Department of State merely requested that American companies refrain from purchasing Libyan oil produced from the nationalized field. Neither in 1972 nor in 1973 (when the companies again tried to boycott Libyan oil) did the American government directly prevent the purchase of expropriated crude by American companies.[42] U. S. officials seem to have considered the Libyan nationalization as a commercial dispute that was then being litigated in American courts.[43]

2. Nationalization in Iraq. The actions of the Libyan government were no threat to the maintenance of the tax floor to price because Libya imposed ceilings on its production and because it negotiated or legislated per-barrel tax increases to the point where its crude became significantly overpriced. However, any government that was willing to cut taxes in order to expand its market share threatened to reduce the revenues of other producers. Therefore, the nationalization of the Iraq Petroleum Company (IPC) on 1 June 1972 followed by the Iraqi government's offer to sell at "reduced and competitive" prices posed a major problem for intra-OPEC diplomacy.

Despite large reserves, Iraqi output had expanded very slowly during 1960-69 because of a continuing dispute between the government and the IPC. During 1972 the IPC cut back Iraqi output because of the slow growth in oil demand. With the lower tanker rates prevailing during 1972, Mediterranean crudes were over-taxed in relation to Persian Gulf crudes, and Mediterranean production fell. The Iraqi government demanded that the operating company either produce at full capacity and deliver the increment in production to the government for direct sale or surrender the Kirkuk oil field to the government. The IPC did not meet either of these terms, and the Kirkuk field was nationalized on 1 June 1972. Unlike

[41]Testimony of G.H.M. Schuler, in *MNC Hearings* 5: 93.

[42]Mira Wilkins, "The Oil Companies in Perspective," *Daedalus* 104 (Fall 1975): 169.

[43]See Testimony of James E. Akins, in *MNC Hearings* 5:22.

the Libyan government, which exploited a short-term transportation crisis, the Iraqi government made its demands at a time when the majors could easily resist such demands by substituting Persian Gulf production for Iraqi oil. Also, Iraq had a minimum of financial reserves. Accordingly, the Iraqi government was determined to market its nationalized oil even if this required substantial tax discounts.

Thereafter, a series of negotiations took place between producing governments. On 3 June 1972, Libya and Kuwait promised to lend money to Iraq: such loans would alleviate the financial pressure on the government to increase the volume of oil production by offering tax discounts. The trade press reported: "Behind the Arab nations' actions . . . lies an offer by Iraq to sell its newly nationalized oil at a cut rate which would have driven down the revenues received by the other countries for their oil."[44] Libya, of course, was directly affected because it was an important Mediterranean producer. Kuwait historically had played the role of the "swing producer," or the marginal producer where the majors would adjust total supply.[45] Increased Iraqi output, therefore, would have resulted in a corresponding reduction of Kuwaiti and Libyan liftings.

After nationalization, Iraq offered the French government a 24 percent share of the nationalized oil (about 280,000 b/d), citing "the just policy by France towards our Arab causes."[46] France had historically sought opportunities to diversify sources of supply in order to be independent of the majors,[47] and accepted the Iraqi offer and also agreed to serve as a mediator in the compensation talks. There were no discounts on the sales of nationalized crude to the CFP, and this helps to explain why the IPC announced that it was willing to forego legal action to reclaim the expropriated crude. The IPC also sought to avoid a confrontation with the Iraqi government because Iraq could have nationalized the remainder of the company's producing assets. A settlement between the IPC and Iraq over compensation was reached in February 1973.

---

[44] Quoted in Adelman, "Is the Oil Shortage Real?," p. 88.

[45] Testimony of Prof. John M. Blain, 25 July 1974, in *MNC Hearings* 9: 220-21.

[46] *MEES*, 16 June 1972, p. 2.

[47] Jean-Marie Chevalier, *The New Oil Stakes*, trans. Ian Rock (London: Allen Lane, 1975), p. 23.

In summary, the Iraqi government had successfully exploited rivalries among the consuming countries—much as Libya had exploited East-West differences—in order to market nationalized crude. Moreover, the aftermath of Iraq's nationalization demonstrated that producing governments could succeed in diplomatic efforts to avoid competitive tax discounts that would undermine market prices.

G. POLITICS AND PRODUCTION LEVELS IN THE PERSIAN GULF

1. Oil and Politics in Saudi Arabia. Changes in the oil policies of the Persian Gulf states—particularly those of Saudi Arabia—were an important development in the 1972-73 period. For both economic and political reasons, Saudi Arabia held the dominant position among Arab producers. For example, in the fall of 1973 Saudi Arabia alone accounted for 40 percent of total Arab production. Moreover, as the traditional center of Islam, it was the acknowledged leader in the conservative camp of the Arab world.[48] Therefore, the Saudi decision to link oil production to the conduct of U. S. policy in the Middle East was significant because such a decision was likely to be emulated by other Arab producers.

During 1972, Arab leaders made at least fifteen public threats to use oil as a political weapon against the United States.[49] Most of these threats emanated from the "radical" regimes of Iraq and Libya. On the other hand, the Saudi Arabian government—at least in its public pronouncements—adhered to its traditional doctrine that "oil and politics do not mix."[50] However, the militancy of the Iraqi and Libyan governments and of the various Palestinian groups had an impact on Saudi political thinking.[51] King Faisal's dedication to the Palestinian cause was strengthened by his determination to see the disputed portions of Jerusalem restored to Arab control.

There was no precipitous shift in Saudi Arabian policy: one can only identify a gradually expanding commitment by the Saudi government to the principle of using oil to influence U. S. policy in

[48] George Lenczowski, "The Oil-Producing Countries," *Daedalus* 104 (Fall 1975): 61.

[49] Uzi B. Arad et al., *1973-1974 Oil Embargo: Lessons and Future Impact* [Hudson Institute Paper HI-2035-P] (Croton-on-Hudson, N. Y., July 1974), p. 1.6. (This report was prepared for the Pentagon Workshop on the Security Implications of Energy, 19 July 1974, Washington, D. C.)

[50] Hanns Maull, *Oil and Influence: The Oil Weapon Examined* [Adelphi Paper No. 117] (London: International Institute for Strategic Studies, 1975), p. 5.

[51] Lenczowski, "Oil-Producing Countries," p. 61.

THE BREAKDOWN OF COMPANY-GOVERNMENT BARGAINING

the Middle East. In April 1973, Oil Minister Yamani informed U. S. Secretary of State Rogers that his government would not permit the expansion of production levels to meet increasing U. S. import demand unless the Americans abandoned their pro-Israeli position. A short time later, King Faisal threatened to freeze production at then-current levels, and in June the king told an Aramco representative that the principle of a freeze in production had his full support.[52] In an August interview on American television, Faisal stated that increases in oil production would be made contingent upon (a) U. S. assistance to Saudi industrial development, (b) additional Saudi needs for oil revenues, and (c) a suitable political atmosphere in the Middle East.[53] In late August the Saudi government announced a maximum 10 percent per annum increase in output. This increase was conditioned upon compliance of Western consuming governments with Arab political demands.

Throughout 1972 and 1973, Saudi officials attempted to influence U. S. government policy by pressuring the representatives of the Aramco consortium. For example, on 3 and 23 May 1973, Aramco officials met with King Faisal. On both of these occasions, Faisal insisted that company executives do everything within their power to change U. S. policy. At the second meeting, the king stated that Saudi Arabia was in danger of being isolated from the Arab world because of the failure of the United States to support Saudi Arabia. The king warned that he would not alienate the entire Arab world, and bluntly told the Aramco representatives: "You will lose everything."[54]

The Saudi position on the use of oil as a political weapon became more radical in mid-1973. This shift in policy also reflected King Faisal's personal commitment to use oil to further the Arab cause. As Oil Minister Yamani told an Aramco representative on 28 August 1973:

> [A]nyone who knows our regime and how it works, realizes that the decision to limit production is made only by one man, i.e. the king, and he makes that decision without asking for anybody's concurrence.[55]

[52] Arad et al., *1973-1974 Oil Embargo*, p. 1.7.

[53] *Ibid.*

[54] Confidential Memorandum of Meetings in Geneva—May 1973, by W.J. McQuinn, 29 May 1973, in *MNC Hearings* 7: 504, Exhibit 7.

[55] Confidential Cable from Ellas (Dhahran) to Johnston (New York), 28 August 1973, in *MNC Hearings* 7: 542, Exhibit 19.

According to Yamani, the king was "one hundred percent determined to effect a change in U. S. policy and to use oil for that purpose."[56] During August 1973 extensive talks took place between Faisal and President Sadat of Egypt during which Sadat is credited with persuading Faisal that the economic interests of the oil producers were in harmony with the moral obligations of the Arab cause. Limitations on oil output were also discussed, and after the meetings Yamani warned Aramco that production might be restricted down to 7 million b/d from the 8.7 million b/d level achieved in August.[57] However, the basic Saudi position was to avoid public mention of any possible punitive actions, such as production cutbacks.[58] The Saudi regime continued to rely upon the political and military support of the United States, and therefore did not want to press the American government too severely,[59] but even Saudi dependence did not prevent Faisal from throwing his support behind production cutbacks after the outbreak of the October Arab-Israeli war.

2. U. S. Policy Response. Aramco executives considered such threats by the Saudi government to be serious,[60] and company officials reported Faisal's demands to top officials in the White House, the Department of State, and the Department of Defense. However, the American officials whom the companies consulted refused to take the Saudi threats seriously. Their response was that the king was "calling wolf when no wolf exists except in his imagination."[61] American officials told the company executives that Saudi Arabia had faced even greater pressure in the past from Nasser to use oil as a political weapon. Therefore, the company executives concluded, "there is little or nothing the U. S. Government can do or will do on an urgent basis to affect the Arab/Israeli issue."[62] Thereafter, although the company representatives continued to communicate to the U. S. government their belief that an embargo or

[56]*Ibid.*

[57]Cable from J. J. Johnston to Aramco Directors, 28 August 1973, in *MNC Hearings* 7: 542, Exhibit 20.

[58]Lenczowski, "Oil-Producing Countries," p. 61.

[59]Walter Laqueur, *Confrontation: The Middle East and World Politics* (New York: Bantam, 1974), pp. 55-56.

[60]Testimony of W. J. McQuinn, 20 June 1974, in *MNC Hearings* 7: 412.

[61]Cable from J. J. Johnston to Aramco Directors, 1 June 1973, in *MNC Hearings* 7: 508, Exhibit 8.

[62]*Ibid.*

cutback was imminent, the attitude of the responsible U. S. officials remained unchanged.[63] Even after the October war began, a Department of State cable to a number of U. S. embassies abroad continued to discount the possibility of a cutback.[64]

In summary, the use of the "oil weapon" was largely unanticipated by U. S. officials.[65] American policymakers repeatedly ignored Arab representations that bargaining on oil production and price involved political—as well as economic—issues. U. S. dependence upon imported oil rose dramatically during 1973, and there was widespread concern in the United States about the security of oil supplies and about the impending "energy crisis." The United States did not react decisively to Arab political demands or to growing dependence upon foreign imports. How does one explain such non-decision?[66]

The policy response by the United States could have taken a number of forms. First, it would have been possible to manage domestic demand for petroleum in order to minimize dependence on Arab sources. Second, it would have been possible to meet Arab political demands, or to provide new incentives to producing states to maintain the flow of oil. Third, it would have been possible to cooperate with other consuming countries in order to counteract the increasing cohesion of the producing governments.

All of these policies were considered. For example, a confidential Department of State report circulated in early 1972 urged a reduction of oil imports through a "crash program" to develop domestic energy sources, and a "major international effort" to promote unity among consuming countries.[67] However, no action was taken to moderate import demand or to accumulate petroleum stocks in the United States. Moreover, OECD meetings designed to establish an oil-sharing program were unsuccessful because of divisions among the

[63] Testimony of George Keller, 20 June 1974, in *MNC Hearings* 7: 429.

[64] U. S. Department of State Confidential Cable no. 202315, 12 October 1973; cited in Robert B. Stobaugh, "The Oil Companies in the Crisis," *Daedalus* 104 (Fall 1975): 185.

[65] "Prior to the OAPEC embargo, the United States government had only the vaguest plans for dealing with such a contingency" (Richard B. Mancke, *Performance of the Federal Energy Office* [Washington, D.C.: American Enterprise Institute, 1975], p. 2).

[66] Peter Bachrach and Morton S. Baratz, "Decisions and Nondecisions: An Analytical Framework," *American Political Science Review* 57 (September 1963): 641-42.

[67] Maffre, "U. S. Oil Interests in Bargaining with OPEC," p. 815.

major consumers on the allocation of supplies.[68] Finally, in 1972 the United States rejected a Saudi proposal to guarantee the flow of Saudi oil to the American market if such oil was admitted duty free and if the Saudi government was permitted to invest in downstream operations within the United States.[69]

There were two basic reasons for the continued inaction of the U. S. government. First, government officials did not realize the extent of the shift of the balance of power in the oil market, and therefore ignored the threats of producing governments. In particular, the experience of the 1967 embargo seems to have convinced officials that Arab threats were not serious.[70] Second, the oil import issue in the United States was a source of contention among a number of powerful domestic political constituencies. Legislation on energy matters was delayed by struggles among federal agencies and among members of Congress.[71] In addition, the government apparatus was poorly designed to produce effective action without strong leadership, and after the spring of 1973 the executive office was increasingly preoccupied with Watergate.

H. CONCLUSION

In our bargaining model, both economic and political factors play a part in the determination of oil prices, and monopoly power—once achieved—can be utilized to attain either political or economic objectives. The increasing monopoly power of the exporting countries in the 1972-73 period was accompanied by attempts of the Arab exporting countries to use oil as a weapon to influence U. S. policy in the Middle East, and this development is consistent with our model.

Despite the repeated threats to use the "oil weapon," the major consuming countries did not react decisively because the implications of changes in the distribution of market power were not understood

---

[68] See Stobaugh, "Oil Companies in the Crisis," p. 185.

[69] The proposal was made by Ahmed Zaki Yamani, "Prospects for Cooperation Between Oil Producers, Marketers and Consumers: The Issue of Participation and After," in Middle East Institute, *World Energy Demands and the Middle East* (2 vols.; Washington, D.C.: Middle East Institute, 1972), 1: 95-105.

[70] See p. 94 below.

[71] James W. McKie, "The United States," *Daedalus* 104 (Fall 1975): 82. See also Richard B. Mancke, "The Genesis of the U. S. Oil Crisis," in *The Energy Crisis and U. S. Foreign Policy*, eds. Joseph S. Szyliowicz and Bard E. O'Neill (New York: Praeger, 1975), pp. 67-72.

THE BREAKDOWN OF COMPANY-GOVERNMENT BARGAINING

by their policymakers. For example, the U. S. Department of State continued to treat "commercial" issues relating to the production and price of oil apart from "political" issues.[72] More important, divisions within the American government and among the governments of the major consuming countries prevented the adoption of policies that might have reduced the likelihood of an embargo or minimized its effects.

The entire course of events from early 1972 to October 1973 may be summarized as follows: The producing countries achieved greater cohesion and moved steadily toward the assertion of total control over supply despite the continued appearance of bilateral negotiations with the companies. The logistic capabilities of the companies were exploited by the producing governments in order to increase revenues, and the companies found that their interest in maintaining their margins was no longer served by resisting tax increases. Had this situation continued, OPEC probably would have evolved during the next several years into an effective governmental cartel. However, the decision of the Arab producers to use the "oil weapon" accelerated this evolution.

---

[72] See Joseph S. Szyliowicz, "The Embargo and U. S. Foreign Policy," in *The Energy Crisis and U. S. Foreign Policy*, pp. 217-19.

PART THREE

THE POLITICS OF OPEC PRICE DETERMINATION

Chapter VI

THE 1973-1974 ARAB OIL EMBARGO

The production cutbacks as well as the embargo imposed by the Organization of Arab Petroleum Exporting Countries (OAPEC) in October 1973 marked a final shift of the power to determine price to the exporting governments. According to an authoritative source, "the successful use of the oil weapon by the Arab states in connection with the Middle East war of October produced the greatest shock, the most potent sense of a new era, of any event of recent years."[1] In the next two chapters, we shall analyze the decision of the Arab producers to use the oil weapon, and shall examine the impact of that decision upon bargaining over price. In this chapter, we will be concerned with the policies of the producing states and the role of the international oil companies; the reactions of the major consuming countries will be analyzed in the next chapter.

A. POSTWAR SUPPLY INTERRUPTIONS

From the early 1940's, the use of oil as a weapon has been a recurring theme in Arab political thought.[2] Supply interruptions had occurred in 1951-53, 1956-57, and 1967. These supply interruptions should be differentiated from the 1973-74 embargo, however, because during them the international oil companies had been able to shift sources of supply to meet demand.[3] Before 1973, excess capacity had always been greater than the shortfall in production created by any producing government.

The 1951-53 interruption of oil production in Iran followed Mossadegh's nationalization of the Anglo-Persian Oil Company. In

---

[1] International Institute for Strategic Studies (IISS), *Strategic Survey 1973* (London: IISS, 1974), p. 1.

[2] Fuad Itayim, "Strengths and Weaknesses of the Oil Weapon," in IISS, *The Middle East and the International System: II. Security and the Energy Crisis* [Adelphi Paper No. 115] (London: IISS, 1975), p. 1.

[3] George T. Patton, "Current Outlook for the Energy Sector," API, Washington, D.C., October 1974, p. 6.

that instance, although Iran was supplying 7 percent of the non-Communist world's crude in 1950, a 95 percent decrease in Iranian production had little impact on total world supply. In response, Middle Eastern production outside of Iran increased by 43 percent in 1951, and Caribbean production increased 13 percent in the same year.[4] Also, in 1950 Iran was supplying almost one quarter of all refined products sold outside of the Western Hemisphere,[5] but the shortfalls were compensated for by larger U. S. product shipments and by increases in refinery runs in Western Europe.

The crisis that occurred in 1956 after the nationalization of the Suez Canal was far more serious than the 1951-53 Iranian shutdown. Following the Anglo-French intervention in Egypt, Nasser closed the canal on 31 October 1956, and it remained closed until the end of March 1967. Three days later, the Syrian army destroyed sections of the IPC's pipeline and thus halted the flow of Iraqi oil to Mediterranean ports. A transport crisis of considerable proportions resulted since the closing of the canal and the rupture of the pipeline directly affected the movement of 2 million b/d.[6] The problem of supplying Western Europe was further aggravated by a lack of available tanker tonnage necessary for the long haul now required around the Cape of Good Hope.[7] The shortfall was covered largely by an increase in U. S. production under the Emergency Oil Lift Program: the equivalent of 73 percent of the decline in oil shipments from the Middle East to Western Europe was compensated for by increased shipments from the United States and from the Caribbean.[8] There was also a considerable drawdown in stocks of refined products, and demand in Europe was held below normal levels by rationing and mild weather.

The third postwar supply interruption in 1967 followed the Arab-Israeli war in June of that year. The Suez Canal, Tapline, and IPC pipeline were all closed on 6 June 1967. The crisis interrupted the flow of some 10 million b/d of Arab oil.[9] In addition to the

[4]Harold Lubell, *Middle East Oil Crises and Western Europe's Energy Supplies* (Baltimore: Johns Hopkins, 1963), pp. 6-7.

[5]U. S. Department of the Interior, Office of Oil and Gas, *Middle East Petroleum Emergency of 1967* (2 vols.; Washington: GPO, 1969), 2: A-1 [Historical Appendix].

[6]Shwadran, *The Middle East, Oil and the Great Powers*, p. 539.

[7]Department of the Interior, *Middle East Petroleum Emergency* 2: A-4 [Historical Appendix].

[8]Lubell, *Middle East Oil Crises*, p. 13.

[9]Department of the Interior, *Middle East Petroleum Emergency* 1: 17.

interruption of supply, Saudi Arabia, Kuwait, Iraq, and Libya on 5 June 1967 halted all exports. This production shutdown, however, continued less than two weeks (except in the case of Iraq, where production was not restored until July). A selective embargo against West Germany, the United States, and Great Britain[10] continued until the end of August, but without production limitations, the embargo proved ineffective.[11] Although the embargo was directed against the United States, only 2.3 percent of total U. S. oil consumption in 1968 was from Arab sources, and increases in domestic production more than covered the shortfall. Iran and Venezuela both took advantage of the embargo to increase their volumes of exports. In Iran the increase of production in 1967 over 1966 was twice the increase of production in 1966 over 1965.[12] By the 29 August summit conference in Khartoum, it was clear that the embargo had hurt only the Arab states, and it was lifted shortly thereafter.[13] During the 1967 emergency, there was a substantial surplus of tanker tonnage; thus the closing of the Suez Canal did not create a major transportation crisis. Since there was no problem with the availability of crude, coping with the 1967 emergency was simply a matter of manipulating the tanker fleet.[14]

The 1967 supply interruption was handled by the industry with a minimum of U. S. government action. The success of the companies in dealing with the 1967 disruption seems to have convinced officials in the consuming countries that the international companies could deal with similar circumstances in the future.[15] For example, an official U. S. government assessment of the 1967 emergency concluded that "in the final analysis, it is the oil companies which actually produce, refine, transport and sell oil to the consumer."[16] Since the 1967 embargo never resulted in a reduction of crude availability, the logic of this assessment is questionable. The author of a 1963 study was more acute when he pointed out that Western complacency concerning the disruption of the flow of Middle Eastern oil was due to the belief that if one producer shut down, oil could easily be

---

[10] Iraq also embargoed Rhodesia and South Africa.

[11] Itayim, "Strengths and Weaknesses of the Oil Weapon," p. 1.

[12] Shwadran, *The Middle East, Oil and the Great Powers*, p. 540.

[13] Stocking, *Middle East Oil*, p. 459.

[14] Department of the Interior, *Middle East Petroleum Emergency* 1: 2.

[15] Shwadran, *The Middle East, Oil and the Great Powers*, p. 544, note 15.

[16] Department of the Interior, *Middle East Petroleum Emergency* 1: 1.

replaced from alternative sources: "It is obvious, however, that the number of alternative sources varies inversely with the number of producing countries that—by design or by accident—do act in concert."[17]

B. THE POLITICS OF EMBARGO

Compared with the abortive 1967 embargo, the economic and political circumstances surrounding the 1973 embargo were strikingly different. By 1973 the bulk of the world's surplus producing capacity was located in the Middle East. Even before the outbreak of the October 1973 war, the major Middle Eastern producers asserted full control over the pricing of crude. The Arab economic sanctions which followed the outbreak of the war further constricted supply and reinforced OPEC's collective power over price.

In September 1973, OPEC members demanded a revision of the Teheran/Tripoli agreements in order to compensate for the declining share represented by government revenues from higher market prices for oil. Negotiations with the companies began in Vienna on 8 October 1973, but were rapidly overtaken by the events following the outbreak of fighting in the Middle East a day earlier.[18] At the Vienna meeting, OPEC members demanded a 100 percent increase in posted prices.[19] Such an increase in taxes would have doubled the cost of crude; the industry negotiators stated they did not have the authority to meet such demands, and asked for a recess to consult with the consuming governments.[20] The real reason for the companies' action was probably a desire to avoid being criticized for "negotiating" such a dramatic increase in per-barrel taxes. The industry realized that there was little they could do to resist the

[17] Lubell, *Middle East Oil Crises*, pp. 17-18.

[18] A more detailed chronology is contained in Federal Energy Administration, Office of International Energy Affairs, *U. S. Oil Companies and the Arab Oil Embargo: The International Allocation of Constricted Supplies*, prepared for the Subcommittee on Multinational Corporations, Committee on Foreign Relations, U. S. Senate, 94th Cong., 1st sess., 1975, pp. 13-17 [Appendix I]; and U. S. Congress, House, Committee on Foreign Affairs, *The United States Oil Shortage and the Arab-Israeli Conflict*, Report of a Study Mission to the Middle East from October 22 to November 3, 1973, 93d Cong., 1st sess., 1973, pp. 11-21 [Part II].

[19] House Committee on Foreign Affairs, *U. S. Oil Shortage and the Arab-Israeli Conflict*, p. 12.

[20] Penrose, "Development of Crisis," p. 49.

imposition of higher taxes, and they could only dissociate themselves from further OPEC action. Negotiations were broken off on 12 October; on 16 October, the Persian Gulf producers met in Kuwait and unilaterally raised posted prices by 70 percent. On 1 January 1974, OPEC raised posted prices further by 130 percent in order to reflect the higher prices received for government-auctioned oil during the embargo period. The impact of OPEC's unilateral pricing was to increase government revenue for the marker crude from $2.00/bbl. on 1 October to $3.43/bbl. on 16 October to $9.31/bbl. on 1 January 1974.[21]

1. Imposition of Production Cutbacks. On 17 October, the Arab oil ministers met in Kuwait and agreed on a program to cut production progressively until the implementation of United Nations Security Council Resolution 242 which called for Israeli withdrawal from the Arab territories occupied after the June 1967 war and for a just settlement of the Palestinian problem. Nine Arab countries[22] agreed to cut oil production by a minimum of 5 percent from September 1973 production levels; they also agreed to announce additional 5 percent cuts each succeeding month, and to allow individual states to impose an embargo on the United States.

Originally Saudi Arabia was a moderating influence on the other OAPEC members attending the 17 October Kuwait meeting. The basic intent of the Saudi government appears to have been to alert world opinion to the gravity of the Arab-Israeli conflict rather than to take punitive action against the United States, upon which the Saudi government relied for political and military support.[23] Thus Saudi Arabia favored adopting a decision on production cutbacks at the 17 October Kuwait meeting while supporting only a recommendation with regard to the embargo.

2. Embargo Policy. Largely because of a change in the Saudi Arabian position, the embargo recommendation soon became a stated policy of the Arab exporters. On 12 October 1973, the executives of the chief Aramco shareholders sent a memorandum to President Nixon warning that further U. S. action in support of Israel

[21] Chandler, *Oil—Prices and Profits*, p. 15, table 2. These figures also account for the effects of new participation terms and revised terms for buy-back oil.

[22] The communique was signed by only nine OAPEC members: Saudi Arabia, Kuwait, Libya, Algeria, Egypt, Syria, Abu Dhabi, Bahrain, and Qatar (Iraq did not sign).

[23] Lenczowski, "Oil-Producing Countries," p. 64.

would result in a major interruption in the flow of oil from Arab producers. This memorandum noted:

> We have been told that the Saudis will impose some cut-back in crude oil production as a result of the United States position taken thus far. A further and much more substantial move will be taken by Saudi Arabia and Kuwait in the event of further evidence of increased U. S. support of the Israeli position.[24]

Company executives told the President that they were "convinced of the seriousness of the intentions of the Saudis and Kuwaitis and that any actions by the U. S. government at this time in terms of increased aid to Israel will have a critical and adverse effect on our relations with the moderate Arab producing countries."[25] Nixon probably never read the memorandum,[26] and U. S. missions in the Arab producing countries were never consulted.[27] However, it is unlikely that either step would have altered American policy.

The Saudi position on embargo policy reversed dramatically after Nixon's 19 October request of $2.2 billion in military assistance for Israel. On 19 and 20 October, Saudi Arabia announced a series of restrictions, including a 10 percent cutback of production—instead of the minimum of 5 percent agreed at the 16 October OAPEC meeting in Kuwait—and an embargo of the United States. These measures were to set the pattern for the sanctions imposed by the other Arab states. The official Saudi declaration of its embargo on oil trade with the United States asserted that "in view of the increase in American military aid to Israel, the Kingdom of Saudi Arabia has decided to halt oil exports to the United States."[28] There is every indication that U. S. action in resupplying Israel also influenced the decision of the other Arab producers to announce an embargo.

There was a conflict within OAPEC over the nature of the Arab economic sanctions. The economic restrictions imposed by OAPEC members from 7 October to 2 November are summarized in Table 5, which indicates substantial differences in the economic sanctions

---

[24] Memorandum to the President, 12 October 1973, in *MNC Hearings* 7: 546-47.

[25] *Ibid.*, p. 547.

[26] Wilkins, "The Oil Companies in Perspective," p. 171.

[27] House Committee on Foreign Affairs, *U. S. Oil Shortage and the Arab-Israeli Conflict*, p. 14.

[28] *Ibid.*

THE 1973-1974 ARAB OIL EMBARGO

Table 5

SANCTIONS IMPOSED BY OAPEC MEMBERS: 6 OCTOBER-2 NOVEMBER 1973

Date	Country	Production Cutback in Percent	Embargo on: U.S.	Embargo on: Netherlands	Other
October					
6	Algeria		X		
7	Iraq				Nationalized Exxon and Mobil assets
18	Saudi Arabia	10%			
	Qatar	10			
	Libya	5			
	Abu Dhabi		X		
19	Libya		X		
	Bahrain	5	X		Cancelled 1971 U.S. naval base agreement
	Saudi Arabia		X		
	Algeria	10			
21	Kuwait	10	X		
	Dubai		X		
	Qatar		X		
	Bahrain		X		
	Algeria			X	
22	Iraq				Nationalized Dutch portion of Shell assets
23	Kuwait			X	
	Abu Dhabi			X	
24	Qatar			X	
25	Oman		X	X	
30	Libya			X	
	Bahrain			X	
November					
2	Saudi Arabia			X	

Sources: FEA, *U. S. Oil Companies and the Arab Oil Embargo*, pp. 13-14 (Appendix 1), and Arad et al., *1973-1974 Oil Embargo*, pp. 1.8-1.9.

imposed. Such differences can be explained by the conflicting interests of the various Arab producers.

At the 16 October meeting in Kuwait, Iraq proposed the nationalization of foreign oil company assets and other assets located in the Arab world belonging to the citizens of unfriendly nations. The official Iraqi position was that a policy of production cutbacks did not distinguish between friendly and hostile consumer states,[29] but the real reason for Iraq's opposition to the OAPEC decision on production limitations was its status as a producer. The dispute with the IPC throughout the 1960's had resulted in the slow growth of Iraqi production; if Iraq accepted the OAPEC program of production cutbacks, Iraq would further compromise future growth of production. Moreover, punitive nationalizations by other producers would lend legitimacy to Iraq's previous actions.[30] Saudi Arabia, in particular, opposed the Iraqi proposal, and the other producers did not favor a policy that threatened both to disrupt production and to invite reprisals from the West. The Iraqi proposal was rejected, and Iraq abstained from all further meetings of the Arab oil ministers during the embargo period. Iraq embargoed the Netherlands and the United States, but it increased exports between September and December by 7.5 percent. Damage to Eastern Mediterranean terminals prevented Iraq from exporting even more oil during the embargo. Its refusal to accept production cutbacks, therefore, had little effect on the total availability of Arab crude.

Iraq was the only defector from the joint Arab position, but there were important differences in objectives among the other exporting countries. The smaller Gulf states—Kuwait, Qatar, Abu Dhabi, and Bahrain—supported the Saudi position on production cutbacks at the 17 October OAPEC meeting, and closely followed Saudi leadership on embargo measures; these states maintained close relations with the Saudi Arabian government and traditionally followed Saudi initiatives.[31] Algeria, although it had a great need for oil revenues for economic development, traditionally was a militant supporter of Arab causes, and it imposed a 10 percent production cutback. Libya had previously used nationalization as a foreign policy instrument, and might have been expected to support the Iraqi position; the Libyan government, however, did not support Iraq. Two reasons suggest themselves for Libya's support of OAPEC

---

[29]*MEES*, 28 December 1973, p. 12.

[30]Lenczowski, "Oil-Producing Countries," pp. 62-63.

[31]*Ibid.*, p. 64.

action: first, by rejecting the cutbacks supported by Egypt and by the vast majority of OAPEC, Libya would have significantly worsened its already strained relations with the Arab world; second, further nationalization of producing assets in Libya would have disrupted production. (In Iraq, nationalization was essentially an accounting operation because Exxon and Mobil were only shareholders in the operating company.)[32]

At the beginning of November, total Arab oil production was reduced by about 20 percent—chiefly because Saudi Arabia and Kuwait first shut-in liftings previously accounted for by embargoed nations and then, in addition, applied the 10 percent cutbacks. Both Saudi Arabia and Kuwait had large financial reserves, and both were concerned with the conservation of national oil assets.[33] In addition, Kuwait's militant position on the production cutbacks may have been influenced by the presence of some 150,000 Palestinians in its population. Libya, on the other hand, continued to produce at levels above those set by OAPEC, and some crude exports still reached the Caribbean for final shipment to U. S. markets during the first month of the embargo.

3. The Evolution of Embargo Policy. The level of Arab economic sanctions was further escalated under Saudi Arabian leadership. Beginning on 21 October, Arab producers extended the embargo to include the Netherlands, chiefly because of pro-Israeli statements by Dutch leaders and because of the use of the Dutch national airline for flights to Israel in connection with its war effort. The inclusion of the important Rotterdam market in the embargo greatly increased its scope. The Saudi government also escalated the nature of the demands that were supported by the economic sanctions. The war in the Middle East ended on 25 October without a military resolution of Arab demands for the return of the occupied territories, and the major exporters announced the embargo would remain in effect. On the same day, an Aramco official reported to his New

[32]*Ibid.*

[33]"[O]il experts with whom the study mission spoke agree that the Saudi's underestimate their reserves and have an extremely conservative view of their production capability" (House Foreign Affairs Committee, *U. S. Oil Shortage and the Arab-Israeli Conflict*, p. 18). In an 18 April 1973 conversation between Yamani and an Aramco executive, the Saudi oil minister indicated official fears that Aramco was overproducing and thus damaging reservoirs (Memorandum to File re Conversation between Moses and Yamani, 19 April 1973, in *MNC Hearings* 7: 560, Exhibit 31). Kuwait imposed limits on production in 1973 to conserve oil resources.

York office that his contacts in the Saudi government had convinced him that Faisal was "radical" on the issue of implementing Resolution 242. He stated that "there is absolutely no question that oil cutbacks will remain in effect until entire implementation of the resolution is worked out."[34] James Akins, the American ambassador to Saudi Arabia, agreed with this assessment, and at Akins's urging the companies communicated the Saudi position to other U. S. government officials.[35]

At the request of the Saudi government, a second OAPEC meeting was held in Kuwait on 4 November 1973. At this time the impact of the embargo was minimal because there was a lag of about one month before the supply restrictions could take effect: tankers required a month to reach Europe from the Persian Gulf. Moreover, the production cutbacks had not deterred the United States from resupplying Israel, and the embargo had no effect on the U. S. airlift of military equipment. The 4 November Kuwait meeting produced a further escalation of economic sanctions as follows: (a) November production would be reduced by a uniform 25 percent below September levels; (b) the 25 percent reduction would include liftings shut-in as a result of embargoes. It was also agreed that Belaid Abdesselam, oil minister of Algeria, and Sheikh Yamani, oil minister of Saudi Arabia, would be sent to visit European capitals in order to explain the Arab actions, after which decisions would be reached as to the countries to be exempted from production cutbacks.

The embargo policy was further refined and coordinated in a series of meetings throughout November and December. At the Arab summit meeting in Algiers on 26-28 November, further changes were made in the "oil weapon." All diplomatic and economic relations—including the export of oil—were severed with South Africa, Rhodesia, and Portugal. Japan and the Philippines were exempted from the December cutback of 10 percent, and it was agreed that non-embargoed European nations would be compensated for the loss of supplies usually shipped through Rotterdam.

These changes in the embargo policy reflected four considerations. The first consideration was an effort to distinguish friendly from hostile consuming countries. For example, following the 6 November EEC declaration calling for a peace settlement which

[34] Cable from J. J. Johnston to Aramco New York Office, 25 October 1973, in *MNC Hearings* 7: 530, Exhibit 16.

[35] Cable from Jungers to Aramco Executives, 26 October 1973, in *MNC Hearings* 7: 517, Exhibit 13.

would conform to UN Security Council Resolution 242,[36] the Arab oil ministers in their meeting of 18 November exempted most EEC states from the scheduled 5 percent December cutbacks. By the end of December, the consuming countries were assigned to one of the following four categories:[37]

(a) *Most favored*—Most favored nations would receive exports to meet current levels of demand. Included the following countries: Britain, France, Spain, the Arab countries which were net oil importers, a number of African countries, and countries which had broken off diplomatic relations with Israel.

(b) *Preferential*—Countries given preferential status would receive September 1973 levels of imports. Included countries which had adopted a pro-Arab position, such as Japan, the Philippines, and Belgium.

(c) *Neutral*—Neutral countries would receive their prevailing level of oil imports less the general level of production cutbacks. Included most EEC countries.

(d) *Embargoed*—Embargoed countries would receive no crude or refined products from Arab exporting countries. Included the United States, the Netherlands, Portugal, and South Africa.

The second consideration which caused changes in embargo policy was the reduction of the demands supported by the embargo. This shift was caused chiefly by changes in the view of the Saudi Arabian government. During Secretary of State Kissinger's November diplomatic effort, which brought about an Egypt-Israeli accord on the management of the ceasefire arrangements, Saudi Arabia gave no sign of softening its position that Israel should withdraw from all occupied territories before the lifting of the embargo and the production cutbacks. However, on 9 December the Saudi government changed its policy and submitted to the Arab oil ministers a proposal that linked the removal of economic sanctions to gradual withdrawal by Israel from the occupied territories.[38]

The third consideration in the changed formulation of Arab policy was the concern of the "conservative" producers not to

[36] Full text of the resolution appears in *The Times* (London), 7 November 1973.

[37] FEA, *U. S. Oil Companies and the Arab Oil Embargo*, p. 17.

[38] Arad et al., *1973-1974 Embargo*, p. 1.11.

pursue production cutbacks to the point where they might cause extensive economic damage to the West and thus provoke economic, or even military, reprisals from the United States.[39] Moreover, King Faisal was strongly anti-Communist, and was concerned with the possible collapse of the European economies and the advantages that such a collapse might give to the Communists.[40]

The fourth and most important consideration in the moderation of Arab demands was that the producing governments had little control over the international distribution system, and they soon realized that they could not enforce a selective embargo. The Arab producers were aware that additional cutbacks would largely punish nations in the most favored or preferential categories, without having any real effect on the American economy. Thus the moderation of demands and the increases in production levels reflected the growing Arab realization of the limitations of oil policy as a selective punitive instrument.[41] As a result, in the meeting of Arab oil ministers in Kuwait on 24-25 January 1973 it was decided that crude production would be raised by 10 percent in January, which would increase output to 85 percent of the pre-embargo level.

In the final stage of de-escalation of sanctions, President Sadat began exerting pressure upon producing governments to terminate the embargo because of the 17 January Egypt-Israeli agreement on the disengagement of troops. While a lifting of the embargo was resisted for some time by Libya, Syria, Algeria, and Saudi Arabia, the embargo against the United States was ended on 18 March 1973. The embargo against the Netherlands was lifted on 11 July.

C. THE ROLE OF THE COMPANIES DURING THE EMBARGO

The international oil companies played the key role in determining the impact of the economic sanctions imposed by the Arab governments. The inability of the OECD nations to reach an oil-sharing agreement relegated the management of the cutbacks and of the embargoes to the companies themselves. The companies decided individually that directives of the Arab governments would be followed. However, they also determined that the reduction of supplies

---

[39] Szyliowicz, "Embargo and U. S. Foreign Policy," p. 205.

[40] Lenczowski, "Oil-Producing Countries," p. 70.

[41] In this connection see Yamani's widely reported remarks that the oil embargo against the United States and the Netherlands had been ineffective (*Washington Post*, 18 January 1974).

would be spread evenly among the major consuming areas. The position taken by Frank McFadzean, chairman of Shell Transport and Trading, was typical:

> We do not regard it as being a function of a multinational enterprise such as ourselves, to start allocating in scarce conditions.... On the question of Arab oil we have complied strictly, as we are bound to do, with the destination controls imposed on the U. S. Where we have got non-destination controlled oil, we will use that to meet the needs of the countries that would otherwise go short. And I don't think an international enterprise such as ourselves has any alternative.[42]

Despite some pressure from the consuming governments, there is every indication that the oil companies adhered to this policy. The experience of Aramco was fairly typical. Yamani informed the Aramco representatives in a 21 October 1973 meeting of the new supply restrictions. The Saudis were aware of the difficulties in administering an embargo, but the company representative reported "they are looking to Aramco to police it."[43] Yamani made it clear that noncompliance by the companies would result in nationalization.[44] The Saudi government also imposed strict controls over the transportation of oil exports in order to prevent diversions to embargoed areas. Thus the major international companies provided the Arab exporting governments with the personnel and facilities to manage the embargo. However, the companies also reallocated supplies among the consuming countries, and effectively undermined the selectivity of the economic sanctions.

The U. S. government did not attempt to prevent American companies from complying with the directives of the Arab governments. American companies immediately notified the Pentagon and the Department of State that they intended to follow the instructions of the producing governments, and they were never told to do otherwise.[45] The U. S. government was unprepared to take

---

[42]Quoted in Peter Hill and Roger Vielvoye, *Energy in Crisis: A Guide to World Oil Supply and Demand and Alternative Sources* (London: Robert Yeatman, 1974), p. 74.

[43]Cable from Jungers to Aramco New York Office, 21 October 1973, in *MNC Hearings* 7: 515, Exhibit 12.

[44]FEA, *U. S. Oil Companies and the Arab Oil Embargo*, p. 7.

[45]Testimony of Otto Miller, 20 June 1974, in *MNC Hearings* 7: 451. See also M. A. Adelman, "How to Have an Oil Crisis—A One-Year-Later Critique," *The Conference Board Record* 12 (January 1975): 44.

Table 6

NON-COMMUNIST WORLD CRUDE OIL PRODUCTION DURING
EMBARGO: SEPTEMBER 1973-MARCH 1974

(in million b/d)

Countries	1973				1974		
	September	October	November	December	January	February	March
OAPEC	20.8	19.8	15.8	16.1	17.6	17.9	18.5
Non-OAPEC	38.4	38.9	39.0	39.3	39.6	39.5	39.5
Total	59.2	58.7	54.8	55.4	57.2	57.4	58.0

Source: Statistics from governmental sources as reported in *Oil and Gas Journal*, various issues, 1973-1974.

decisive action, and it apparently accepted the companies' position that their refusal to follow the directives of the exporting governments would have only resulted in additional shortages of crude. As an Aramco director observed later: "Obviously it was in the best interests of the United States to move 5, 6, 7 million barrels a day to our friends around the world rather than to have that cut off."[46] The effect of the Arab cutbacks on non-Communist world crude oil production is summarized in Table 6.

The Arab production cut of 5 million b/d in November (of which 2.3 million b/d was Saudi production) was slightly offset by increases in Canadian, Iranian, and Indonesian production.[47] Thus the lowest monthly rate of crude production during the embargo was only 7 percent below the September pre-embargo level. However, the shortfall of oil moving in international trade was a quite considerable 14 percent. Moreover, the embargo shortfall was even more dramatic in light of the 11 percent average per annum growth rate in international oil trade during the previous five years.[48]

The decision of the companies to "equalize the suffering" in allocating production cutbacks among the major consuming areas,

[46] Testimony of George Keller, 20 June 1974, in *MNC Hearings* 7: 418.

[47] Stobaugh, "Oil Companies in the Crisis," p. 180.

[48] *BP Statistical Review 1973*, p. 21.

and thereby to ignore the categories established by the Arab oil ministers, apparently seemed to be the safest course of action in order to minimize the possibility of legislation or litigation by the consuming governments. In addition, this course of action preserved the political neutrality of the companies. Finally, it maintained the commercial role of the companies—that is, to provide flexibility of supply from different sources.[49]

The United States did not pressure the companies to increase American imports at the expense of other countries. The head of the FEA testified before the Senate that the U. S. government urged the international companies to "bring as much as possible into the United States," but at the same time to recognize the "interests of all countries in the world in having some kind of equitable share of the world's supplies."[50] Some consuming countries, however, pressured the companies in order to obtain higher oil imports during the embargo. For example, Prime Minister Edward Heath met with executives of Shell and BP on 21 October 1973, seeking to ensure preferred treatment for the United Kingdom by pointing out that his country was in the Arab most-favored category.[51] Heath was told that Britain would be treated like any other country, and would suffer an immediate 10 percent cut in supplies. Heath became extremely angry, but company executives merely told him that Britain should enact legislation requiring company compliance with Heath's demands. Britain's relations with its European allies would not have been served by such legislation, and Heath acquiesced in the cutback. French and Italian officials also informed the oil companies that they expected deliveries of crude oil at normal levels,[52] but this did not prevent the majors from announcing a 10 percent cutback in supplies for December.

In order to make the necessary reallocations of oil, the international companies were forced to alter about 30 percent of their normal distribution pattern. They significantly increased the flow of unembargoed oil to the United States, redirected embargoed oil to non-embargoed European nations, and relied largely upon Shell to supply oil to the Netherlands.[53] However, a number of restrictions

---

[49] Stobaugh, "Oil Companies in the Crisis," p. 187.

[50] Testimony of Dr. John C. Sawhill, 5 June 1974, in *MNC Hearings* 9: 3.

[51] An account of the incident is reported in Sampson, *Seven Sisters*, pp. 262-63.

[52] *PIW*, 5 December 1973, p. 1.

[53] FEA, *U. S. Oil Companies and the Arab Oil Embargo*, p. 9.

placed on oil trade by consuming countries made this redistribution more difficult; for example: Italy and Spain restricted exports of refined products, the Netherlands halted exports to the United States, and Belgium required export licenses for refined products.

The companies managed to circumvent such restrictions and maintain a relatively equitable distribution of available supplies.[54] The companies were not without considerable bargaining power in negotiations with the consuming nations: if product exports were restricted by consuming governments, crude oil could be shipped to another refinery in another country.[55] Moreover, the companies actually resorted to an embargo of their own in order to obtain higher product prices. In March 1974, Petrofina of Belgium joined ten other companies in refusing to import crude into Belgium unless higher prices for refined products were permitted by the government.[56]

Table 7 indicates the performance of the companies in managing the shortfall in supply during the embargo period. The figures are for the five U. S. majors, but are generally representative. Western Europe and the United States received roughly equal treatment; Japan received most favored treatment, and its imports actually increased by one percent during the embargo period. In general, the companies made allowances for differences in the growth rates in oil demand by using an historical base-period or by considering the projected growth in demand. When adjustments are made for the different growth rates in national energy demand, the allocation of supplies appears more equitable. The FEA calculation of the distribution of shortages as a percentage of projected current demand for oil during the embargo period shows that the percentages were roughly uniform: (a) the United States showed a -17.0 percent change; (b) Europe showed a -18.6 percent change; and (c) Japan showed a -16.0 percent change.[57] Thus the allocation plan of the companies resulted in an equitable distribution of supplies. There is no evidence that the companies systematically overcharged during the embargo, or favored markets where higher margins were obtainable.[58]

[54] See statement of R. B. Stobaugh, 25 July 1974, in *MNC Hearings* 9: 189-90.

[55] Stobaugh, "Oil Companies in the Crisis," p. 191.

[56] *PIW*, 18 March 1974, p. 4.

[57] FEA, *U. S. Oil Companies and the Arab Oil Embargo*, p. 9.

[58] Statement of R. B. Stobaugh, in *MNC Hearings* 9: 188.

THE 1973-1974 ARAB OIL EMBARGO

Table 7

COMPARATIVE DELIVERIES OF CRUDE OIL AND PRODUCTS TO THE LARGEST CONSUMING AREAS BY FIVE U. S. MAJOR OIL COMPANIES DURING A BASE-PERIOD AND THE EMBARGO[a]

(In thousand b/d)

Consuming Area	Base-Period: December 1972- March 1973	Embargo: December 1973- March 1974	Change in Percent
United States	2,607.3	2,294.8	-12.0%
Canada	506.1	474.6	- 6.3
Western Europe	4,859.3	4,198.3	-13.1
Japan	1,683.3	1,701.0	+ 1.0
Other	2,795.3	2,918.9	+ 4.4
Total	12,451.3	11,586.8	- 6.9

Source: FEA, *U. S. Oil Companies and the Arab Oil Embargo*, p. 11, table 2.

[a] Oil companies are Exxon, Gulf, Mobil, Socal, and Texaco.

The real impact of the embargo, however, cannot be measured by using a base-period in which the price of crude was almost four times lower. The posted price of crude increased by 70 percent and then by 130 percent during the embargo period. *The Economist* checked insurance coverage at Lloyds in order to measure tanker movements and found that "in the peak boycott month of January 1974 shipments of oil from the Middle East were 5 percent higher than in the free market of January 1975."[59] Since the price levels in these two periods were roughly equivalent, *The Economist* concluded that during the embargo the Arab exporters were shipping more oil than consumers were ever likely to purchase at the new higher prices.[60] This may be somewhat of an overstatement in view of (a) the lack of information concerning crude availability during the embargo, (b) the high demand for stocks in early 1974, and (c) a further downturn in economic activity in 1975. The conclusion of

[59] *The Economist*, 8 March 1975, pp. 16-17.
[60] *Ibid.*

The Economist suggests, however, that the most important aspect of OAPEC action was not the embargo itself, but the production cutbacks and their effect on prices.

D. THE PRICE OF CRUDE DURING THE EMBARGO

In this section we shall briefly examine the conflict among the producing countries over the pricing of crude during the embargo period. In their 22 December 1973 meeting in Teheran, the Persian Gulf producers raised posted prices by an additional 130 percent effective 1 January 1974. This step increased government revenue on the marker crude to $7/bbl. The OPEC countries also established a simplified pricing system using Arabian light 34° API as the marker in accordance with which the prices of other crudes were fixed, with standard adjustments for quality and transport differentials. This was an important step because it removed the likelihood of concealed discounts on taxes.

The Iranian government played the dominant role in the 22 December Teheran meeting. It both proposed and did much to secure the dramatic increase in government revenue. The Shah's desires to develop his country and to have Iran play a major role in the Persian Gulf required significant financial resources. Moreover, Iran had a high rate of production in relation to reserves, so it was less concerned than other producers with the long-run impact of higher prices upon the development of substitutes for Middle Eastern oil.[61] The Shah attempted to justify the dramatic increase in per-barrel taxes by citing the high cost of alternative sources of energy.[62]

Prior to the Teheran meeting, seasonal demand and Arab production cutbacks had resulted in a dramatic rise in spot prices for crude. In early November it was reported that there were 200 bidders for a single lot of Nigerian oil.[63] In such a chaotic market, competitive overbidding was a certainty, and cargoes of Algerian and Nigerian crude were sold at prices of $16/bbl. In December the National Iranian Oil Company (NIOC) received the staggering price of $17.40/bbl. The companies repeatedly complained to representatives of the producing governments about the effects of the price increase. Company executives strongly suggested to OPEC that no changes

---

[61] Maull, *Oil and Influence*, p. 24.

[62] See Sampson, *Seven Sisters*, p. 257.

[63] House Foreign Affairs Committee, *The U. S. Oil Shortage and Arab-Israeli Conflict*, p. 17.

in per-barrel taxes should be made until market conditions stabilized.

On the other hand, the Shah intervened directly in the oil ministers' meeting in Teheran on 28 December and proposed a $14/bbl. posted price. Kuwait and Iraq were in favor of the Iranian proposal, but Saudi Arabia opposed the proposal and was the decisive factor in moderating the price increase. The final posted price which was set at the meeting was $11.65/bbl. for 34° API crude—a compromise between Yamani's suggested $7.50/bbl. and the Shah's proposal of $14/bbl.

Saudi Arabia opposed larger tax increases because of Faisal's fears that—if carried too far—higher prices might cause extensive damage to Western economies. He also believed that the increase in taxes would diminish the "political" significance of the embargo. Long-run revenue maximizing objectives were not an important determinant of the embargo pricing policy of Saudi Arabia. In the last analysis, Saudi Arabia compromised on higher prices because the absence of such compromise would destroy OPEC. Furthermore, Saudi Arabia could prevent the implementation of higher prices only by significantly expanding its own production, which would force the Saudis to abandon the very economic sanctions designed by them.

E. CONCLUSION

The October 1973 war provided a catalyst for OAPEC solidarity that was necessary for the Arab countries to impose joint economic sanctions effectively. The Arab sanctions, in turn, created the market conditions in which OAPEC asserted complete control over price determination. OAPEC cohesion was also favored by two additional factors. First, there was a lack of surplus producing capacity outside of the Middle East. This meant that the Arab countries did not have to reduce production below 25 percent, and did not have to absorb actual losses in revenue. Second, Arab production was concentrated in Kuwait and Saudi Arabia. These two countries accounted for 56 percent of all Arab oil moving in world trade in September 1973, and both countries gave strong support to the policy of production limitations to achieve political ends.

A number of conclusions emerge from the preceding analysis of the objectives and implementation of the "oil weapon":

First, the embargo imposed by the Arab countries was not a

selective instrument because the oil companies controlled the international distribution system. As the *Petroleum Economist* pointed out:

> [T]he successful use of weapons of war rests on the ability to distinguish friend and foe and to ensure that strategic gains are reaped by the user. The oil weapon, by its very nature, cannot do this.[64]

Second, the logistic capabilities of the international companies served the consumers well when their governments were unable to agree upon a formula to distribute oil among themselves. By equitably allocating constricted oil supplies, the companies preserved their long-range objectives to be as free as possible from consuming government restrictions, and to maintain their reputation for dependability. They wisely eschewed a strategy of short-term profit maximization.

Third, neither the threat nor the actual implementation of an embargo had any appreciable effect upon U. S. policy in the Middle East.[65] The use of the "oil weapon" may have added an urgency to efforts by the United States to promote a peace settlement. Even if the "oil weapon" had not been used, the United States would, in all probability, have taken the same steps in order to defuse a major threat to world peace.[66] The Arab production cutbacks posed a considerably more serious problem for Europe and Japan than they did for the United States. At the time of the embargo the United States was relying on Arab sources for no more than 10 percent of its total supplies. By contrast, European dependence upon imports was 70-80 percent of total supplies.[67] Nevertheless, Europe and Japan could not play an important role in the Middle East conflict.

Finally, the Arab economic sanctions transformed OPEC into an effective cartel. The production cutbacks directly influenced market prices; they also resulted in the imposition of additional per-barrel taxes and in a huge increase in producing-government

[64]*PE*, December 1973, p. 442.

[65]The opposite view is expressed in Itayim, "Strengths and Weaknesses of the Oil Weapon," p. 4. He concludes that "the Arab oil embargo was probably the major cause of the change towards a more realistic American policy in the Middle East. . . ."

[66]Joseph A. Yager and Eleanor B. Steinberg, *Energy and U. S. Foreign Policy* (Cambridge, Mass.: Ballinger, 1975), p. 33.

[67]*PE*, November 1973, p. 402.

THE 1973-1974 ARAB OIL EMBARGO

revenue. It is quite unlikely that Arab producers would have undertaken such substantial cutbacks in production in the absence of the October war. Thus the Middle East conflict was responsible for the rapid increases in the market price of crude, but the maintenance of higher market prices required restrictions on output. After the lifting of the embargo, OPEC faced the problem of determining how such restrictions in supply were to be distributed among its members.

Chapter VII

INTERNATIONAL POLITICS AND THE PRICE OF CRUDE, 1973-1975

In this chapter we shall analyze the policies adopted by the governments of the consuming countries in the OECD area in response to the imposition by the exporting governments of economic sanctions and higher per-barrel taxes. During this period, the consuming governments generally took the line of least resistance and paid a higher monopoly rent to the oil exporters.[1] We are concerned here only with the policies of the consuming governments that were designed to affect the production and price of crude oil from Middle Eastern sources. The broader issues of managing international economic adjustment in light of the oil price increase are outside the scope of our analysis.

A. ENERGY POLICY IN THE OECD AREA BEFORE OCTOBER 1973

As shown in Chapter II above,[2] Western Europe had witnessed a steady decline in the coal industry during the postwar period. By 1972 energy consumption was heavily oriented to imported oil. Coal output was maintained only by large government subsidies to the industry.[3] Previous interruptions in oil supply had influenced European policies: in most countries there were mandatory stocking requirements. During 1973, European countries had about ninety days of oil and products in storage and another thirty days of oil in shipment.[4] There were also coal stocks on hand to meet about eighty

---

[1] See Statement of Arthur M. Okun, 5 March 1975, in U. S. Congress, House, Committee on Ways and Means, *The Energy Crisis and Proposed Solutions*, 94th Cong., 1st sess., 1975, 1: 297-98.

[2] See p. 11.

[3] Richard L. Gordon, *The Evolution of Energy Policy in Western Europe: The Reluctant Retreat from Coal* (New York: Praeger, 1970), p. 319. For a review of coal protectionism, see also Adelman, *World Petroleum Market*, pp. 277-78.

[4] *MEES*, 19 October 1973, p. 4.

days of consumption. These stockpiling practices were more extensive than in the United States and in Japan, although in 1973 stocks of refined products in the United States equalled seventy days of imports.

In discussions on European oil and energy policy throughout the 1960's and early 1970's, one is struck by the frequent appearance of such phrases as "security" and "assurance of supply." Although "security" and "assurance of supply" could be obtained at the price of maintaining stockpiles of crude and products,[5] the use of these terms in Europe has been linked primarily to distrust of the large—and mostly American—international oil companies.[6]

Such distrust found expression in the policies of the French and Italian governments. The French government had licensed importing and refining since 1926; also, the CFP sought access to equity crude throughout the 1960's in order to achieve independence from the majors.[7] In Italy, ENI pursued policies that reflected a similar motivation: the Italian government was uneasy about leaving the supply and distribution of this strategic commodity in the hands of foreign companies.[8] In the early 1970's, Japanese firms—with government backing—concluded contracts directly with producing governments for crude oil.

It should be emphasized that considerable effort and expense in the consuming countries was directed toward minimizing reliance upon the majors. Many European countries sought to ensure that the United States and Great Britain would not be able to manipulate the supply policies of "their" oil companies. There was no real danger; the 1967 embargo demonstrated the unlikelihood of such an occurrence. The 1973 embargo proved that the majors were not likely to serve as effective instruments of any nation: the multinational oil companies escaped control by any single government by being subject to the power of all governments.[9]

[5]Gordon, *Evolution of Energy Policy in Western Europe*, pp. 314-16.

[6]Adelman, *World Petroleum Market*, p. 246.

[7]For example, Madelin concludes that the establishment of state-owned companies, such as the CFP, Enterprise de Recherches et d'Activités Pétroliers (ERAP), and Ente Nazionale Indrocarburi (ENI), were national reactions to the undue influence held by the majors (Henri Madelin, *Oil and Politics*, trans. Margaret Totman [Lexington, Mass.: D. C. Heath, 1975], pp. 140-41).

[8]Marcello Colitti, "Vertical Integration, Major Oil Companies and Newcomers: The Case of ENI," paper presented at the Oil Seminar, St. Antony's College, Oxford, 2 March 1976, p. 10.

[9]Edith Penrose, "Multinationals: Dilemma for Individual Countries," *The Times* (London), 22 February 1974.

Despite the experience of the 1967 emergency, European and Japanese leaders continued to identify U. S. policy in the Middle East and the predominance of the "American" majors as the chief threats to the security of their oil supplies. For example, many European and Japanese leaders believed that it was American oil imports which drove up prices for Middle Eastern crude before October 1973,[10] and the embargo was considered a direct result of U. S. policy. Such perceptions help explain European and Japanese fears at the time of the 1973 embargo that the companies were favoring the United States and were trying to maximize profits by withholding oil from markets where product prices were controlled.

On the other hand, before the 1973 embargo there was comparatively little concern in Europe about the possibility of interruptions in supply or of unilateral price determination by the producing governments. Policymakers apparently believed that the role of the United States in maintaining a balance of power between Israel and the Arab states made the United States vulnerable to Arab economic sanctions. They did not believe that Europe was in the same position. Thus, policymakers in the OECD area believed during the 1973 embargo that dissociating themselves from the United States would make oil supplies more secure. Cooperation on oil policy among the major importing nations, therefore, became even more difficult to achieve after the embargo and production cutbacks were announced.

The final aspect of oil policy in the OECD area that we shall consider is management of oil import demand. The U. S. government did nothing to prohibit competitive bidding by American firms. Japan, on the other hand, briefly attempted to control the competitive bidding, which had raised market prices for crude in 1973. After Japan Lines signed a contract with the UAE in February 1973 for participation crude at prices above the going market rate, the Japanese Ministry of International Trade and Industry took steps to limit bidding by Japanese firms.[11] However, the policy was abandoned shortly thereafter, and the Japanese government did not prevent Japanese trading firms, utility companies, and independent refiners from bidding up prices at auctions of crude oil.[12]

The problem of controlling competitive bidding was only one symptom of the general failure of the consuming countries to coor-

[10]Uzi Arad, "The Dilemmas of Interdependence: Western Cooperative Options in the Face of the Oil Challenge," *Middle East Information Series* 26-27 (Spring/Summer 1974): 59.

[11]*MEES*, 9 February 1973, p. 2.

[12]Yoshi Tsurumi, "Japan," *Daedalus* 104 (Fall 1975): 125.

dinate their oil policies. For example, Japan ceased its controls over bidding because of a conviction that the U. S. and European governments would not (or could not) do the same.[13] OECD countries found it difficult to coordinate their oil policies because of important differences in national energy policies and objectives. For example, Europe and Japan were far more dependent upon oil from Middle Eastern sources than was the United States. Thus, maintaining an uninterrupted flow of oil imports was a policy objective of paramount importance for these countries. On the other hand, the United States, as a world power with global objectives, accorded a higher priority to the maintenance of freedom of action in the international arena than to the maintenance of the flow of imported oil from Middle Eastern sources. Even within Europe, there were problems in coordinating various national approaches to oil policy, which ranged from the French *dirigisme* to the German free-market doctrine.[14]

B. POLICY RESPONSE IN THE OECD AREA

1. Western Europe and Japan. After the outbreak of the October war and the declaration of Arab economic sanctions, Western European governments took a neutral position. With the exception of the Netherlands, all refused to brand Egypt or Syria as violators of the 1967 truce. France had maintained a consistent pro-Arab stance since 1967, and Britain had placed an embargo on arms shipments that adversely affected Israeli military capabilities. After the Middle East cease-fire, the West German government requested American authorities to cease using military facilities in West Germany for arms shipments to Israel. These policies were designed to minimize the chance of conflict with the Soviet Union, and to assure continued access to oil supplies from Arab sources.

The Dutch government, in response to the OPEC embargo against it, asked the EEC to share oil resources. This request, however, interfered with British and French desires to avoid a confrontation with Arab producers. The Dutch request precipitated a "vulnerable state of indecision" in European governments.[15] In addition, the British government reportedly feared that if it shared oil supplies with the Netherlands, it might face pressure in the future to share its

[13]*Ibid.*
[14]Romano Prodi and Alberto Clô, "Europe," *Daedalus* 104 (Fall 1975): 96.
[15]*Ibid.*, p. 98.

indigenous oil and gas resources with other EEC countries.[16] President Pompidou of France noted that the Dutch had previously resisted community energy proposals because they were a threat to Rotterdam's position as Europe's largest oil port.[17] The Dutch realized that the embargo posed a similar threat, and would thereby benefit Antwerp, Dunkerque, Le Havre, and Fos.

During the first weeks of the embargo, Europe pursued three main policies:

First, European countries imposed a number of measures to limit consumption—for example, speed limits, driving restrictions, gasoline rationing, and restrictions upon heating and lighting.[18]

Second, European governments attempted to limit the export of refined products in order to ensure accommodation of their domestic markets. Thus Belgium, the Netherlands, Italy, Spain, and Great Britain all imposed forms of export controls, but the international oil companies rerouted oil and avoided such controls in their attempts to equalize the shortfall in supply.[19]

Third, European countries actively attempted to avoid identification with U. S. policy in the Middle East, and sought to be classified by the Arabs as "preferential" countries. Accordingly, EEC foreign ministers reached a common position on the Middle East conflict in a resolution adopted on 6 November 1973. Their resolution endorsed UN Resolution 242 and called for an end to Israel's territorial occupation and for consideration of the "legitimate rights" of the Palestinian people. A member of the community's information commission noted:

"Community foreign ministers were regarded as tacitly acknowledging the power of the Arab oil weapon. Their bid to 'appease' Arab governments gave rise to fears that they had made themselves vulnerable to further pressure from the Arab oil producers."[20]

---

[16]Martin U. Mauthner, "The Politics of Energy," *Middle East Information Series* 26-27 (Spring/Summer 1974): 63.

[17]*Ibid.*

[18]Hill and Vielvoye, *Energy in Crisis*, pp. 77-88, and Joseph A. Yager and Eleanor B. Steinberg, "Trends in the International Oil Market," in *Higher Oil Prices and the World Economy*, eds. Edward R. Fried and Charles L. Schultze (Washington, D. C.: Brookings Institution, 1975), p. 242.

[19]See pp. 109-10 above.

[20]Mauthner, "Politics of Energy," *Middle East Information Series* 26-27: 64.

European policies reflected the concern of governments with obtaining "access" to crude. If product prices had been allowed to increase, the market itself would have rationed available oil, but instead the governments relied upon restrictions. Even in mid-December, when European ports were experiencing difficulties in handling arriving tankers and storage space was at a premium, European governments were announcing the need for new conservation measures.[21]

In order to design a common course of action to deal with the supply interruption, a summit meeting was held by government leaders in Copenhagen in mid-December. At the meeting a conflict developed between the Benelux countries, which requested joint action, and Britain and France, which were concerned with maintaining their "most favored" status.[22] As a result, all that emerged from the summit was an agreement to share statistical information, and even this agreement was not honored.[23]

Japanese policy during the embargo was quite similar to that of the European countries. Prior to the 1973 embargo, Japan had followed a more pro-Arab policy than any other large consuming country with the exception of France,[24] but Japan was accused of "odious neutrality" by the Arab exporters and was subjected to the production cutbacks.[25] The OAPEC classification of Japan as "unfriendly" to the Arab cause struck Japan as "a thunderbolt from the blue sky," in the words of the Japanese Ministry of Foreign Affairs.[26] Japanese officials were advised by Secretary of State Kissinger not to capitulate to OAPEC demands. However, after the 18 November 1973 announcement that the EEC countries would be exempted from OAPEC December production cutbacks in exchange for their endorsement of the Arab position, the Japanese business community put increasing pressure on their government to take similar action.[27] On 22 November 1973, the Japanese government "clari-

[21] *The Economist*, 15 December 1973, pp. 73-74.

[22] Prodi and Clô, "Europe," p. 104.

[23] *Platt's Oilgram News Service*, 7 January 1974, p. 2.

[24] M. A. Adelman, "Politics, Economics and World Oil," *American Economic Review* 64 (May 1974): 62.

[25] *New York Times*, 18 October 1973.

[26] Tsurumi, "Japan," p. 124. We should point out that such public announcements of embargo policy were not always implemented. For example, a Shell executive told me that "Shell for one never received any official evidence that Japan had been classified as 'unfriendly.' We supplied Arab oil to Japan throughout the embargo period, and this was never queried by any producing country."

[27] *Ibid.*

fied" its position on the Middle Eastern conflict in a statement that was regarded as favorable to the Arab position. Japan was subsequently reclassified by OAPEC as a "neutral" country. This Japanese announcement was the first open rupture in postwar diplomatic relations between the United States and Japan. The Japanese government also imposed a series of fuel-saving measures designed to curb consumption by 10-12 percent,[28] and restricted the export of refined petroleum products.

In summary, the importing countries in the OECD area that were most dependent on Arab oil pursued policies that left some latitude for national maneuvering. Considering the variety of energy policies in the OECD area, the inability of the consuming governments to develop an oil-sharing arrangement during the supply crisis is hardly surprising. Also, the internal political difficulties of European countries made dramatic policy initiatives unlikely. Between late 1973 and mid-1974, administrations changed in Britain, France, Germany, and Belgium. Thus, declarations of position on the Middle East conflict (on which Europe and Japan could have little influence) seemed to consuming governments an expedient means of ensuring the continued flow of vital oil supplies.

2. The United States. Most of the American policies adopted in response to Arab economic sanctions were primarily designed to regulate the domestic allocation of crude oil. Hence the chief function of the Federal Energy Office (FEO) was to spread the burden of the petroleum shortages within the United States in an equitable and efficient manner.[29] Like Western European governments, the U. S. government put the heaviest restrictions upon consumer uses of petroleum.[30] Despite the emphasis of the FEO on the domestic allocation of oil, one aspect of FEO policy had a direct impact on U. S. import demand by limiting competitive bidding in international markets. On 1 February 1974, the FEO ordered American refiners with crude oil supplies in excess of the industry average to sell some of

[28] Hill and Vielvoye, *Energy in Crisis*, p. 95.

[29] Mancke, *Performance of the Federal Energy Office*, pp. 7-8. The FEO was organized immediately after the embargo, and was later replaced by the Federal Energy Administration (FEA).

[30] "Government policy, in attempting to minimize the impact of shortages, was designed to accommodate industrial needs and to promote conservation in the private consumption uses of energy" (FEA, Office of Economic Impact, "The Economic Impact of the Oil Embargo on the American Economy" [Washington: FEA, 8 August 1974], p. 3).

their "surplus crude" to crude-poor competitors. Since domestically produced crude was price-controlled, crude-poor refiners could reduce their costs by buying "surplus crude" for several dollars less per-barrel than the cost of an equivalent foreign crude. Likewise, crude-rich refiners would lose several dollars per-barrel on oil imports which exceeded the industry average. As a result of this restriction, U. S. demand for imports declined.[31] By the time the restriction was implemented, however, the competitive bidding had already had its principle effect in provoking OPEC demands for higher per-barrel revenues.

U. S. foreign policy in the Middle East did not respond immediately to the Arab economic sanctions. After the outbreak of the October war, the United States concentrated its efforts on resupplying Israel and on securing a cease-fire and disengagement. Later, during Kissinger's November 1973 "shuttle diplomacy," he failed to persuade King Faisal to lift the embargo.[32] Although Faisal sought to avoid a confrontation with the United States, his position in the Arab world would have been jeopardized by abandoning the sanctions to which he had committed himself only weeks earlier.

After the Egyptian-Israeli cease-fire and the exchange of prisoners, Kissinger stated at a 21 November 1973 press conference that "economic pressures" against the United States were "inappropriate" during the peace negotiations, and that "it is clear that if present measures continue unreasonably and indefinitely, then the United States will have to consider what countermeasures it will have to take."[33] Kissinger's threats were interpreted as implying U. S. military intervention, and produced a strong reaction from Arab leaders. The next day Sheikh Yamani warned that any military action would force the exporting governments to sabotage production facilities and would result in the collapse of the European and Japanese economies.[34] In an interview in early January 1974, Kissinger again asserted that military action would be considered in the case of "some actual strangulation of the industrial world."[35] Kissinger's threats of military action were accompanied by a number of calculated "leaks"

---

[31] Mancke, *Performance of the Federal Energy Office*, p. 15.

[32] Szyliowicz, "Embargo and U. S. Foreign Policy," p. 206.

[33] *New York Times*, 22 November 1973.

[34] *Ibid.*, 23 November 1973.

[35] *Business Week*, 13 January 1974, p. 67.

concerning the training of the U. S. marines for the invasion of desert countries.[36]

Professor Knorr has pointed out that the reasons for not launching an American military response to the Arab "oil weapon" were numerous: military intervention might have provoked a political reaction in the United States; it would have threatened the position of the Persian Gulf leaders friendly to the United States, namely Faisal and the Shah; it might have led to a complete shutdown of oil production; and finally it might have even resulted in Soviet military intervention.[37] The public statements by U. S. officials were probably designed to reinforce private warnings that there was a limit to the economic disruption that would be tolerated by the United States. King Faisal, however, was also concerned about the economic consequences of the "oil weapon."

In January 1974, U. S. attention shifted from access to crude to the more fundamental issues of oil pricing and production. For example, on 31 January 1974 Kissinger told a Congressional committee that far more important than the lifting of the embargo were the issues of adequate future production and the stabilization of current prices.[38] Accordingly, Kissinger intensified efforts to organize a joint response to higher prices by the major consuming countries. But most European countries and Japan attempted to increase their individual bargaining power by seeking better bilateral relations with oil producing governments.

C. BILATERAL AND MULTILATERAL RESPONSES TO HIGHER OIL PRICES

The events immediately following the outbreak of the October war were not conducive to cooperation among the United States, Western Europe, and Japan. The United States resupplied Israel without prior consultation with its NATO allies while the EEC issued its 6 November 1973 resolution calling for Israeli withdrawal without consulting the United States.[39] A similar statement by the Japanese government strained Japanese-American relations. There was no coordination among OECD countries concerning oil conservation meas-

---

[36]I. F. Stone, "War for Oil?," *New York Review of Books*, 6 February 1975, p. 7.

[37]Knorr, "The Limits of Economic and Military Power," p. 237.

[38]*New York Times*, 1 February 1974.

[39]McKie, "The United States," p. 86.

ures, restrictions on the export of petroleum products, or the sharing of oil.[40] Prior to the embargo, an oil-sharing system had been discussed within the OECD, but these discussions did not produce an accord: agreement could not be reached on the treatment of U. S. domestic production within an oil-sharing plan.

In early November, Britain and France, seeking to secure crude supplies, initiated a series of high-level discussions with the oil producing states.[41] U. S. officials considered these negotiations irresponsible, and Kissinger pressed for a common stand by the consuming countries. Accordingly, at American request a conference was held in Washington in February 1974, and included representatives of the EEC, Canada, Japan, and Norway. The consuming countries that relied heavily on imported oil did not want to identify themselves with any agreement which resembled an "oil buyers' cartel." In fact, the French officially characterized the meeting as an "exchange of views," and the final communique referred merely to a comprehensive program to deal with "all facets" of the world energy situation. The Washington conference established an Energy Coordinating Group that was instructed to prepare detailed plans for future meetings.

The U. S. government's strategy throughout 1974 consisted of four main elements, as follows:

First, the United States pressed members of the Energy Coordinating Group to establish an oil-sharing plan that would protect the consuming countries against another embargo.

Second, the United States sought an international agreement to enforce oil-sharing measures and to develop alternative sources of energy; such an agreement would augment the impact on the world market of the U. S. plans for energy independence. (The basic assumption of American policy was that OPEC countries would not be able to maintain price discipline if the demand for OPEC oil was dramatically reduced.)

Third, the United States attempted to discourage other consuming countries from trying to make bilateral agreements with oil producing governments.

Fourth, the United States sought to improve its relations with

[40]Ulf Lantzke, "The OECD and Its International Energy Agency," *Daedalus* 104 (Fall 1975): 220.
[41]*New York Times*, 14 February 1974.

Saudi Arabia and to convince the Saudi government to expand oil production.[42]

The United States completely failed to achieve the last two objectives. Despite American objections, France, Italy, Britain, and West Germany early in the embargo period made direct arrangements with producing governments.[43] In January and February 1974, France made agreements to exchange oil for industrial equipment and technical assistance with Iran, Libya, and Saudi Arabia; in February 1974, Italy made similar agreements with Libya and Saudi Arabia; in January 1974, the Federal Republic of Germany agreed to build a petrochemical complex in Iran that would be financed by natural gas and oil sales; in the same month, the United Kingdom agreed with Iran to exchange crude oil for industrial goods.[44] The price paid for oil in these agreements is difficult to assess because of the "barter" elements in the transactions, but it is generally believed that oil was priced at close to tax-paid cost.[45] (In other words, the consuming countries did not obtain any substantial discounts.) On balance, such bilateral agreements appear to have had relatively little impact in enhancing the bargaining power of the consuming countries, and 90 percent of the oil exported from OPEC countries in 1974 continued to be handled by the international companies.[46]

One result of these bilateral agreements has been the acquisition by the oil exporting countries of sophisticated weapons systems;[47] a significant percentage of the estimated $32.5 billion spent by OPEC countries on imports during 1974 was used for the purchase of military hardware. In 1974, the United States alone sold arms worth $4.4 billion to Persian Gulf countries; Britain, the Soviet Union, and France supplied military equipment worth a further $3.5 billion—according to one estimate.[48] Such sales enhanced Iran's military

---

[42]Henri Simonet, "Energy and the Future of Europe," *Foreign Affairs* 53 (April 1975): 454. See also Kissinger's address at the University of Chicago on 14 November 1974; summarized in *PE*, December 1974, p. 454.

[43]*PE*, February 1974, p. 51.

[44]For a more detailed summary, see Stockholm Institute for Peace Research, *Oil and Security* (New York: Humanities Press, 1974), pp. 116-17.

[45]*PE*, March 1974, p. 83.

[46]See p. 138 below.

[47]Edward M. Kennedy, "The Persian Gulf: Arms Race or Arms Control," *Foreign Affairs* 54 (October 1975): 25.

[48]Susan Hart, "The Disposition of OPEC Revenues," in British-North America Committee, *Higher Oil Prices: Worldwide Financial Implications* (Washington, D. C.: British-North America Committee, 1975), pp. 25-26 [Appendix].

capabilities relative to the other Persian Gulf countries.[49] Although the rapid arms buildup in the Persian Gulf may influence oil pricing in the near future,[50] it does not explain 1973-75 price developments. At most, one can say only that the availability of weapons—coupled with Iran's security program—increased Iran's need for revenues, and perhaps made the Shah more militant on the issue of increasing per-barrel revenues.

An important aspect of U. S. policy was to improve bilateral relations with the Saudi government. Therefore, soon after the lifting of the embargo, the U. S. government entered into an agreement to strengthen Saudi security forces. The Department of State, which had assiduously criticized European and Japanese bilateral agreements with producers, denied that this agreement was similar,[51] but there were suspicions among Arab leaders that the primary function of the Saudi-American agreement was to outflank European and Japanese efforts to secure bilateral oil deals,[52] and the agreement demonstrated to European and Japanese leaders that the United States could dominate a continuing scramble for bilateral agreements.

The U. S. strategy of promoting multilateral measures for oil-sharing and oil conservation finally produced results in November 1974, when agreement was reached on the establishment of an autonomous International Energy Agency (IEA)[53] set up within the OECD framework. In essence, the agreement provided for (a) an emergency sharing arrangement that would allocate oil among the consuming countries in the case of an embargo, (b) a long-term cooperative program designed to reduce dependence on OPEC oil, (c) an exchange of information on the world oil market, and (d) a framework for coordinating negotiations with the oil-producing governments.[54]

---

[49]Dale R. Tahtinen, *Arms in the Persian Gulf* (Washington, D. C.: American Enterprise Institute, 1974), p. 30.

[50]The staff of the Hudson Institute has constructed a number of scenarios relating to possible interruptions of oil supplies because of military conflict among the Persian Gulf countries; see Arad et al., *1973-1974 Oil Embargo*, pp. 5.1-5.15.

[51]Szyliowicz, "Embargo and U. S. Foreign Policy," p. 207.

[52]*MEES*, 12 April 1974, p. 6.

[53]Agreement on International Energy Program, in U. S. Congress, House, Committee on International Relations, Subcommittee on International Organization and on International Resources, Food and Energy, *Legislation on the International Energy Agency*, Hearings, 94th Cong., 1st sess., 1975, pp. 56-79.

[54]Statement of Julius Katz, 18 December 1974, in U. S. Congress, House, Committee on Foreign Affairs, Subcommittees on International Organizations

The most important achievement of the International Energy Program was the emergency oil-sharing accord. The consuming countries agreed—in the event of a future interruption in supply—to allocate oil as the companies had during the embargo in 1973. The agreement also specified minimum levels of emergency stocks.[55] One State Department official described this emergency program as "a short-term insurance policy. It does not in itself deal with the problem of excessive dependence on imported oil."[56] To challenge OPEC's control over oil pricing, the consuming countries would have had to take measures to minimize import demand, but there was no agreement on such measures. During the remainder of 1974, conservation and the development of alternative sources were merely discussed. In these discussions, the United States maintained that sufficient investments in high-cost alternative sources of energy would be made only if these investments were protected. Thus, attention began to be focused upon a common minimum price—maintained by national tariffs or quotas—below which the price of imported oil would not be allowed to fall.

Since there were differences in (a) oil import dependence, (b) the potential for investment in alternative energy sources, and (c) the costs of alternative sources among the consuming countries, no agreement on a common minimum price was reached.[57] In February 1975, the IEA announced a guideline for total imports of 20 million b/d for 1975, but there was no means of enforcing this guideline.

The IEA was a step toward closer cooperation among the consuming countries, but it fell far short of organizing the collective market power necessary to reassert control over the price of crude oil. A further step toward cooperation among consuming countries was taken in February 1975 when Europe and the United States agreed on the establishment of an International Monetary Fund $6 billion reserve fund that would allow OECD countries to finance the balance-of-payments deficits caused by higher oil prices. The OECD nations also agreed in principle to a $25 billion "solidarity fund" that would permit less fortunate nations to borrow from the wealth-

and Movements and on Foreign Economic Policy, *U. S. Policy and the International Energy Agency*, Hearings, 93rd Cong., 2nd sess., 1975, p. 11.

[55]These were set initially at the general levels prevailing before the embargo; see *PE*, November 1974, p. 403.

[56]Statement of Thomas O. Enders, 26 March 1975, in House Committee on International Relations, *Legislation on the International Energy Agency*, p. 8.

[57]*The Economist*, 22 February 1975, p. 12.

ier OECD members in order to finance oil deficits.[58] These agreements to "recycle" OPEC revenues were made possible only by a reversal in U. S. policy. The United States had taken the position that any international action that would allow OPEC governments to invest their surplus revenues safely would encourage higher oil prices, but in the fall of 1974 American officials abandoned their diplomatic efforts to convince the major exporters to lower oil prices and accepted "recycling" measures, hoping that they might pave the way toward greater cooperation on the limitation of OECD oil imports.

D. CONCLUSION

Once OPEC had established full control over the price of crude in October 1973, the consuming countries were unable to exert even a small measure of control over price determination; more specifically, the major oil companies were unable to instigate any collective action. Western Europe and Japan were critically dependent upon imported oil, and were unwilling to risk a confrontation with the exporting governments; instead, they sought to improve their bargaining positions, and attempted to assure crude supplies by a series of bilateral accords with the exporting governments.

Throughout 1974, the efforts of the United States to lower the price of crude oil were largely unsuccessful.[59] The multilateral aspect of American strategy was to promote OECD measures on oil-saving: the United States assumed that such measures would create a surplus of oil and thus break OPEC price discipline. This aspect of American strategy failed because of fundamental differences in dependence upon imported oil among the consuming countries.[60] At the same time, the United States made available a number of political and financial incentives to the Saudi Arabian government in an effort to persuade it to lower its prices for crude. The Saudis did not lower their prices, and the price of oil in the period October 1973-January 1975 continued to reflect directly the collective decisions of the exporting governments.

---

[58]*PE*, February 1975, p. 43; *The Economist*, 15 February 1975, p. 62.

[59]Simonet, "Energy and the Future of Europe," p. 455.

[60]Arad et al., *Energy and Security: Implications for American Policy* [Hudson Institute Report HI-1884/2-RR] (Croton-on-Hudson, N. Y.: Hudson Institute, July 1974), pp. 3-5.

Chapter VIII

INTRA-OPEC BARGAINING AND THE PRICE OF MIDDLE EASTERN CRUDE, 1973-1975

In October 1973 the producing governments asserted full control over the pricing of crude oil. However, their control over crude oil prices rested upon their collective power to regulate production, and hence upon their ability to agree on the distribution of output. In this chapter we shall analyze bargaining over crude oil prices among OPEC members. The first section of the chapter examines OPEC as an international organization; the second section analyzes the structure of the petroleum market in order to determine the economic constraints on the pricing policy of OPEC members; the third section analyzes price developments during the period in terms of bargaining among the major exporting governments.

A. OPEC: STRUCTURE AND FUNCTION

Despite repeated efforts, OPEC never evolved into the governmental cartel on which the organization was modelled. For example, Venezuela championed the cause of prorationing during the 1960's, but the OPEC members did not even adhere to the "transitory" production program established in 1965, which set guidelines for the growth rates of national oil production. Thus the program was allowed to lapse in June 1967.[1] After June 1967 the Economic Commission of OPEC continued to consider possible production controls, but no formal system of prorationing was adopted.

The OPEC organization is essentially an instrument of its member governments: its low level of institutionalization results from the reluctance of the exporting governments to endow it with a measure of autonomy.[2] OPEC decisions can be made only by the unanimous vote of its full membership. For most of its history the organization

---

[1] Schurr and Homan, *Middle Eastern Oil and the Western World*, p. 124. See also Francisco R. Parra, "OPEC: Present and Future Role," in *Continuity and Change in the World Oil Industry*, eds. Mikdashi et al., p. 135.

[2] Mikdashi, "Cooperation among Oil Exporting Countries," p. 26.

had statutory limitations that prevented the evolution of a strong secretariat; until June 1970 the secretary-general's term of office was limited to one year.[3] Also, the brevity of the terms of OPEC staff appointments, and the fact that OPEC personnel are almost always on loan from member governments, creates deficiencies in staffing.[4] Producing governments are generally unwilling to assign their most qualified personnel to OPEC, and the primary loyalties of OPEC staff members are to their national governments. For these reasons, the OPEC secretariat has never been able to equal the expertise of the various member governments, and it has served basically as a clearing house for information and a mechanism to coordinate negotiations.

OPEC provides a forum in which producing governments can consult and plan joint negotiating strategies. Once the member governments have determined that it is in their interest to cooperate, OPEC provides an instrument for cooperative action; however, the history of the prorationing issue demonstrates that when the member governments fail to reach a consensus, OPEC is powerless to act.[5] The most important actions taken by the producing governments during the 1970-73 negotiations were taken outside the OPEC organization. For example, the Teheran accord was negotiated by a committee of Persian Gulf states, the 16 October 1973 unilateral posting of price was undertaken by the Persian Gulf states, and the production cutbacks imposed in Kuwait on 17 October 1973 were the results of decisions by the ministers of the Arab countries under the auspices of OAPEC.

B. THE POLITICAL ECONOMY OF THE OPEC PROCESS

Since its foundation in 1960, OPEC has been concerned almost exclusively with per-barrel revenues, and after the OPEC countries asserted full control over the pricing of crude oil in 1973, bargaining

[3] Fuad Rouhani, *A History of O. P. E. C.* (New York: Praeger, 1971), p. 124. The secretary-general's term of office was increased to two years in 1970.

[4] Zuhayr Mikdashi, "The OPEC Process," *Daedalus* 104 (Fall 1975): 207.

[5] At OPEC's 13 September 1974 meeting held in Vienna, another in a long series of studies of possible production controls was requested from the secretary-general (*PE*, October 1974, p. 375). Continuing Saudi Arabian resistance to prorationing was the factor holding up adoption of such controls, since "no coordinated production cutbacks can be operated without the consent of Saudi Arabia, which represents 28% of the total OPEC output [in 1974]" (*Arab Oil and Gas*, 1 March 1975, p. 6).

within OPEC continued to center around the level of per-barrel taxes upon the marker crude; in turn, the tax-paid cost of crude was used as an instrument to control output. An economic model of the market for OPEC oil might be defined as follows: (a) buyers face a perfectly elastic supply curve at the administered price, and (b) shifts in demand are translated into fluctuations in the quantities of the various crude oils purchased and produced.[6]

1. Output Policy in a Producers' Cartel. Even when price is used as an instrument to control production, our bargaining model suggests that the level of output and the distribution of market shares will reflect the objectives and bargaining power of individual exporting governments. Therefore, we must first analyze a number of economic factors that, according to our model, underlie the policies and bargaining positions of the individual producing governments[7] — namely, (a) the elasticity of demand for OPEC oil in consuming countries, (b) the number of exporters and their share in the market, (c) the relative demand of producing governments for oil revenues, and (d) the potential for a producing government to increase its market share and revenue.[8]

The first factor that might be expected to influence the policies of OPEC exporters is the elasticity of demand for OPEC oil. This factor depends upon the strength of a number of substitution effects against higher priced oil, including (a) substitution of other energy sources for OPEC oil, (b) substitution of other factors of production for higher priced oil, and (c) a rise in price of energy intensive goods relative to non-energy intensive goods, and a decline in consumer purchases of the former. In addition, there is an income effect of higher priced oil: in the absence of compensatory government policy, a rise in oil prices will cause oil imports to fall because real income will decrease.[9] (The income effect works both ways: an increase in real income because of an upturn in industrial activity will increase demand for energy.)

[6]Robert Mabro, "OPEC after the Oil Revolution," *Millennium* 4 (Winter 1975-1976): 193.

[7]See p. 7 above.

[8]See U. S. Congress, House, Committee on Banking and Currency, Ad Hoc Committee on the Domestic and International Monetary Effects of Energy and Other Natural Resource Pricing, *Meeting America's Resource Needs: Problems and Policies*, Report, 93d Cong., 2d sess., 1974, p. 79.

[9]W. M. Corden, "Implications of the Oil Price Rise," *Journal of World Trade Law* 8 (March-April 1974): 133.

This theoretical analysis is almost impossible to quantify. Estimates of the elasticity of oil demand, based upon differing assumptions about the elasticity of substitution, vary widely.[10] Estimates of the elasticity of demand for OPEC oil over time were extremely varied. Nobody had any idea what the demand for $11/bbl. oil would be, compared with that for $2/bbl. oil, until the price increased. The tax-paid cost of the marker crude was $1.70/bbl. in June 1973; by October 1974 it had risen to $9.69/bbl., but despite this dramatic increase in price and the imposition of the embargo, total OAPEC production decreased by about one percent[11] against a forecast increase of about 7.5 percent.[12]

After the downturn in economic activity in 1974 and the reduced demand for stocks of petroleum and products following the termination of the embargo,[13] OPEC exports declined. OPEC output decreased steadily from May 1974 (31.8 million b/d) until April 1975 (25.8 million b/d). What proportion of this decline in oil exports reflected the high income elasticity of demand? There was no radical change in the structure of industrial plant or in the pattern of energy consumption, and important alternative sources of energy did not come onstream. We can therefore conclude that the most important factor affecting the demand for oil was the general level of economic activity.

Let us briefly analyze the producing governments as revenue maximizers in terms of the economic theory of imperfect competition. Considered as an aggregate, OPEC faced the classic problem of any profit-maximizing monopolist—that of equating marginal cost with marginal revenue over time.[14] Since oil is an exhaustible resource, marginal cost includes the costs of oil production and the current value of income foregone by present production. The current value of income foregone depends not only upon the rate of discount but also upon the current price, since the current price will affect the rate of substitution of other energy sources for OPEC oil. (Here it is

---

[10]Most estimates of the short-term elasticity of demand for oil are in the -0.1 to -0.5 range (Economist Intelligence Unit, *Multinational Business*, April 1975, p. 16). Conservation without the development of alternative sources might make demand even less elastic (see M. A. Adelman, "U. S. Energy Policy," in *No Time to Confuse*, p. 35).

[11]Field, "Oil in the Middle East," p. 88 [Statistical section].

[12]*PE*, March 1975, p. 84.

[13]Economist Intelligence Unit, *Multinational Business*, April 1975, p. 14.

[14]Hollis B. Chenery, "Restructuring the World Economy," *Foreign Affairs* 53 (January 1975): 251.

assumed that consuming countries would rely on their own high-cost energy sources, once such sources were developed, even if OPEC in the future lowered its price.)

The foregoing model is often invoked to explain OPEC behavior, but we submit that it is the wrong model. OPEC is not a supranational authority, and therefore a discussion of overall output policy for the cartel is at best a theoretical abstraction. Our objective is to explain the behavior of producing governments in making decisions on price and output during 1973-75, and in our judgment it was the perceptions of the leaders of such governments that were the most important element in such decisions.

It is doubtful that the handful of leaders who determined petroleum policy in the Middle East thought in terms of the long-run elasticity of substitution. While it is true that Iranian demands for a $14/bbl. posted price in December 1973 were supported by arguments based upon the cost of alternative sources of energy, such demands primarily reflected Iran's need for oil revenues to fulfill the Shah's ambitious plans. Iran would have been unwilling to undertake the production restriction which a $14/bbl. posted price would have necessitated.[15] Saudi Arabia was the chief opponent of Iranian demands for higher taxes in December 1973. It could be argued that the large Saudi reserves have made the Saudi government more sensitive to the implications of higher prices for the substitution of other energy sources, but the evidence suggests that Saudi policy was motivated primarily by concern over the economic disruptions that higher prices would have caused.

2. Output Policies of Individual Producers. The price and output policies of all of the OPEC members are interdependent. Thus the effects of the output policy adopted by any individual OPEC member depend not only upon market reactions, but also upon the policies pursued by other OPEC members. Therefore, of the four economic factors that we analyzed with reference to our bargaining model,[16] the three factors relating to individual producing countries are the most important in determining output policy—namely, (a) the market shares of the individual exporters, (b) the revenue demands of the individual exporters, and (c) the potential for an individual exporter to increase its market share.

In Table 8, we have divided the OPEC countries into three

[15] Yager and Steinberg, *Energy and U. S. Foreign Policy*, pp. 66-67.
[16] See p. 132 above.

groups according to similarities in oil production, population, and oil reserves in 1973/74. The countries in Group I had large reserves, small populations, and limited prospects for economic development. For example, the Arab Persian Gulf producers in Group I in 1973 accounted for 65 percent of proven reserves and for 48 percent of OPEC output. Accordingly, the producers in Group I might have been expected to favor lower current price and thereby higher future revenues because their resources other than oil were limited.

The countries in Group II had smaller oil reserves than those in Group I, much larger populations, and good prospects for economic development. Accordingly, the producing countries in Group II might have been expected to favor higher current prices because of the high potential for investment of oil revenues in their own economies. However, the bargaining power of Group II within OPEC was much less than that of Group I because the countries in Group II had lower installed capacity and could not cut back substantially on production because of their serious need for revenues.

The countries in Group III had the smallest reserves, largest populations, and most serious need of revenues for development. Accordingly, the countries in Group III might have been expected to favor very high current prices, but the bargaining power of the countries in Group III was even less than that of the countries in Group II—for similar reasons.

As is shown in Table 8, Saudi Arabia had the highest productive capacity, the largest reserves, and minimal revenue needs. Because of these characteristics, Saudi Arabia was endowed with preponderant bargaining power within OPEC: it could use its large installed capacity to decrease market prices, and it had the flexibility to reduce output dramatically for the purpose of increasing prices because of its minimal needs for revenues. We noted that the countries in Group I (including Saudi Arabia) might have been expected to favor lower current prices, but—as we suggest below—Saudi Arabian production policy was motivated more by political designs than by the economic considerations which were common to Group I countries.

We should emphasize that the different economic circumstances of OPEC countries were not the only source of rivalry within the organization. For example, there were important differences in East-West political leanings among OPEC countries. In addition, there were serious ruptures in diplomatic relations among Arab producing countries because of territorial and political disputes—i.e., Iraq vs. Kuwait, Iran vs. Libya, and Iran vs. Iraq[17]—and there is evidence

[17] Mikdashi, "The OPEC Process," p. 209.

Table 8

SELECTED OPEC COUNTRIES' OIL MARKET SHARES, SHARE OF TOTAL OPEC RESERVES, AND POPULATION IN 1973 AND USABLE PRODUCTIVE CAPACITY IN 1974

Group[a] and Country	Percent Share of OPEC Output: 1973	Percent Share of OPEC Reserves: 1973	Population: 1973 (millions)	Oil Production: 1973 (million b/d)	Usable Productive Capacity: 1974 (million b/d)
GROUP I					
Saudi Arabia	24.8%	33.6%	8.1	7.5	9.7
Libya	7.1	6.5	2.1	2.1	3.0
Kuwait	9.7	16.3	0.9	3.0	3.8
UAE	4.3	5.5	2.9	1.2	2.3
Qatar	1.9	1.7	0.9	0.5	0.7
GROUP II					
Iran	20.1	15.3	31.9	5.9	6.5
Venezuela	11.3	3.6	11.3	3.5	6.5
Iraq	6.3	7.9	10.4	2.0	2.5
Algeria	3.5	1.9	14.7	1.0	1.1
GROUP III					
Nigeria	6.7	5.1	73.4	2.0	2.4
Indonesia	4.4	2.7	125.0	1.3	1.5

Sources: Columns 1 and 2: Bohi and Russell, *U. S. Energy Policy,* p. 62, Table 3.4; Columns 3 and 4, and classification of countries: Chenery, "Restructuring the World Economy," p. 250, Table 1; Column 5: Arnold E. Safer and Anne Parker Mills, *Outlook for World Oil: Prices and Petrodollars* (New York: Irving Trust Company, 20 March 1975), p. 3.

[a] For criteria used for groupings, see pp. 134-35 of text.

to suggest that Saudi political rivalry with Iran concerned not only divergent oil price objectives but also Saudi fears of Iranian designs to control the Persian Gulf.[18]

3. Market Structure and the Output Policy of Individual Producers. In addition to the other factors discussed, bargaining among oil producers over price was influenced by the structure of the oil market. Because of vertical integration and an awareness of each other's investment plans, the major oil companies were able to adjust demand to supply at higher-than-competitive prices, but after October 1973 the majors became mere agents for implementing the decisions on price and supply that the OPEC countries had taken.[19]

Two characteristics of the market in crude oil facilitated price discipline by OPEC members. First, the system of posted prices—which amounted to per-barrel excise taxes—was a clear record of compliance or noncompliance of individual producers with OPEC-wide pricing policy. Second, the large, vertically integrated international companies provided a mechanism for rapidly translating changes in the demand for products in the consuming countries directly into changes in rates of production, which removed the temptation to discount on price in order to move "surplus" oil at a given price structure.[20]

With the rapid movement toward 100 percent participation in 1973-75, there was widespread speculation that OPEC countries would play an increasing role in marketing their own crude, and that this would undermine price discipline. This speculation was based upon the assumption that it would be difficult to prevent price cutting in a market consisting of a large number of competitive buyers many of whom were not vertically integrated.[21] Almost the same argument was made at the onset of the participation negotiations, but the producing governments continued to rely upon the capabilities of the international companies—as they did in the 1973-

---

[18] An Aramco representative was told by a top political advisor to Faisal that increased Iranian military expenditures were for the purpose of obtaining control over the Persian Gulf (Cable from Jungers to Aramco New York Office, 2 May 1973, in *MNC Hearings* 7: 507).

[19] Paul H. Frankel, "Organizationsprobleme der Ölimportländer," in Friedrich-Ebert-Stiftung, *Zukunftsorientierte Energie- und Rohstoffpolitik* (Bonn-Bad Godesberg: Verlag Neue Gesellschaft, 1976), p. 41.

[20] The basic assumption here is that a competitive market with a number of brokers would result in a higher aggregate demand for stocks of oil.

[21] See Adelman, "Politics, Economics and World Oil," pp. 63-64.

75 period. In 1974, for example, 90 percent of oil exports from the Middle East moved through traditional company channels.[22]

In 1973 the majors still retained access to equity crude and could obtain oil at a tax-paid cost that was lower than going third-party market rates. However, the majors' margins were progressively trimmed during 1973-75, and by the end of 1974 the majors' margins were effectively fixed by producing governments.[23]

C. BARGAINING AND THE PRICE OF CRUDE

Bargaining within OPEC was over the price of the marker crude. OPEC did not directly administer prices in order to account for differentials (in quality and transport) among the crudes produced by various countries. Such price differentials were fixed by individual producing governments, and had an important effect upon the distribution of output. Changes in government revenue for the marker crude in the period October 1973-January 1975 are summarized in Table 9;[24] these changes in government revenue were due to conflicts among the OPEC countries.

Throughout the embargo period, the Saudi government opposed any increase in taxes. At its 17-19 January 1974 meeting, OPEC agreed to a three-month freeze on tax-paid cost of the marker crude. In this meeting, Algeria, Indonesia, and Iran (with a serious need for additional oil revenues) proposed a further 15 percent increase. In turn, Yamani threatened to break with OPEC by unilaterally lowering Saudi prices and expanding production. Thus the decision to freeze government revenues was basically dictated to OPEC by Saudi Arabia. In March the embargo was lifted, and there was a decision to restore OAPEC production levels to those prevailing in September 1973. Saudi Arabia, however, raised its output above September 1973 levels, thereby depressing market prices. Thus, auction sales of participation crude in March (the UAE) and in August (Kuwait) fell short of postings and were cancelled.

At the OPEC meeting held in Quito in June 1974, Yamani formally proposed that the posted price for the marker crude be lowered by $2.50/bbl. to $6.81/bbl. Saudi Arabian officials again

---

[22] Parra, Ramos, and Parra, *International Crude Oil and Product Prices*, 15 October 1974, p. ix.

[23] *PE*, March 1975, p. 91.

[24] Since tax-paid cost varied from company to company, government revenue is the best indicator of average price.

Table 9

GOVERNMENT REVENUE PER-BARREL FOR SAUDI ARABIAN 34° API MARKER CRUDE: OCTOBER 1973-JANUARY 1975

Date	Dollars per-Barrel[a]	Explanation of Change
1 October 1973	$ 2.00	
14 October 1973	3.45	The first unilateral posting and revised terms for buy-back oil (retroactive adjustment).
1 January 1974	9.31	The second unilateral posting and new participation terms.
1 July 1974	9.41	Revised terms for buy-back oil and royalty (Quito meeting).
1 October 1974	9.74	Revised terms for tax, royalty, and buy-back oil (Vienna meeting).
1 November 1974	10.14	Revised terms for tax, royalty, and third unilateral posting (Abu Dhabi meeting).
1 January 1975	10.14	Introduction of single price structure (Vienna meeting).

Sources: Chandler, *Oil—Prices and Profits*, p. 15, Table 2, and Field, "Oil in the Middle East," pp. 90-91.

[a] Assumes operating costs of 10 cents/bbl.

threatened to take unilateral action to lower prices by an auction of participation crude without a floor price if Saudi terms were not accepted. It was decided to increase the royalty on equity oil from 12.5 to 14.5 percent, which substantially decreased the profitability of the crude owned by the majors.

At the OPEC meetings, the attitudes of the "price militants" were shaped by their needs for additional revenues: Algeria and Iran, for example, consistently supported tax increases. (Strong "conservationist" producers such as Kuwait and Libya supported increased taxes because they desired to cut back on output.) The "price militants" did not play a controlling role within OPEC in determining the level of taxes, however: only Saudi Arabia had the installed capacity that enabled it to strongly influence the prevailing price structure. The dominant position of Saudi Arabia in terms of productive capacity helps to explain Saudi opposition to prorationing, since a fixed formula for market shares would give the Saudi government less latitude to manipulate production to serve its own objectives.

What were the basic motivations for Saudi production decisions? We suggest that Saudi production policy was influenced primarily by broad foreign policy objectives. For example, in the period immediately after the termination of the embargo, Saudi Arabia made a number of agreements to expand economic, military, and political ties with the United States.[25] At this time, the Saudi government apparently desired to strengthen its traditionally close relations with the United States. The embargo notwithstanding, Aramco remained an all-American concern, and Saudi Arabia continued to maintain a strong anti-Communist position. It would appear that the Saudis issued threats to expand production unilaterally and to lower prices as part of this desire for cooperation with the United States. The Saudi position on lowering prices also won it support in the developing world, where the oil price increase had severely damaged development prospects.

In late 1974 there was a shift in Saudi Arabian policy. It no longer demanded revenue freezes, and Saudi officials only expressed disagreement with OPEC moves toward higher per-barrel revenue. However, the Saudi government began to absorb production cutbacks in order to maintain market prices. This change in policy seems to have reflected a shift in Saudi concerns from political security to the maintenance of OPEC. The new Saudi policy found expression at the OPEC meeting in Vienna, where Saudi Arabia publicly stated its dissatisfaction with the OPEC decision to increase taxes and royalties by 3.5 percent. The official Saudi position was that "increase in average government take is only justified on the basis of excess profits realized by the international oil companies."[26]

Producing government revenue was increased in 1973-75 by the participation agreements—usually on a 60 percent basis—that were concluded toward mid-1974. These agreements increased per-barrel revenues, but they distorted the complicated pricing system, allocating a large share of monopoly rent to the majors that retained equity crude.[27]

---

[25] See p. 127 above.

[26] Oded Remba, "The OPEC Cartel and Oil Prices: Illusions and Realities," *Middle East Information Series*, No. 1 (January 1975), p. 5.

[27] Ian Seymour, "Towards a New Structure for the Oil Industry," Supplement to *MEES*, 15 November 1974, p. 1.

INTRA-OPEC BARGAINING AND THE PRICE OF CRUDE

OPEC meetings were held in September and November 1974 for the purpose of correcting these distortions in the pricing system. Saudi Arabia took the position at the 12-13 September meeting in Vienna that an increase in government revenue through higher taxes on equity crude should be accompanied by a decrease in posted prices, which was consistent with their official policy of reducing "excess profits" of the majors. Other producing governments, however, refused to accept a decrease in posted prices,[28] and the final decision of the OPEC meeting was to increase average government revenue by 3.5 percent through increases in the tax rate on equity crude.

As a result of their dissatisfaction with the outcome of the Vienna meeting, Saudi Arabia—together with the UAE and Qatar—on 10 November 1974 acted independently of OPEC and increased the effective tax rate on equity crude even further than the 3.5 percent specified in the Vienna OPEC decision, while lowering posted prices by some 40 cents/bbl. Although Saudi officials talked of a "price reduction,"[29] the net effect of the Saudi measures was to increase average government revenue by 40-52 cents/bbl. One of the motivations for the Saudi initiative of 10 November was to pressure Aramco and other operating companies into speedy acceptance of 100 percent participation terms. Saudi Arabia also wanted to make the gesture of reducing prices—even though at the same time they increased taxes and government revenues. Finally, the Saudis desired to confront OPEC members with a fait accompli at their next meeting on the price issue.[30]

By late 1974, the shift in policy orientation of Saudi Arabia from efforts to lower prices to maintaining OPEC-wide price discipline was complete. The final stage in this shift was signalled by the indefinite postponement of a scheduled auction of participation crude without a floor price and by the retroactive application of earlier OPEC decisions on higher taxes and royalties. At a meeting in Vienna on 13-14 December 1974, OPEC adopted the pricing system instituted by Saudi Arabia, the UAE, and Qatar during the preceding month. This system established a general crude acquisition cost which narrowed the difference between the tax-paid cost for the

---

[28] *Ibid.*, p. 2.

[29] See Remba, "The OPEC Cartel," *Middle East Information Series*, No. 1, pp. 5-6.

[30] *Ibid.*, p. 6, and Seymour, "Towards a New Structure for the Oil Industry," Supplement to *MEES*, 15 November 1974, pp. 1-3.

operating companies and the Saudi state oil company's sales price to 21 cents/bbl.—the companies' margin.[31] The adoption of this system was an important step in rationalizing the general OPEC price structure, and thereby preventing variations in the revenues of individual countries caused by differences in participation agreements. It produced an increase in government revenues from the higher taxes and royalties, which was followed by a nine-month freeze by OPEC on increases in government revenues.

Saudi Arabia's commitment to the maintenance of OPEC price discipline, as well as its key role as the price leader, is demonstrated by an analysis of the distribution of production cutbacks in the post-embargo period. Such cutbacks were necessary because the aggregate demand for petroleum fell in response to higher prices and slower economic activity in 1974. Consequently, the capacity utilization of processing equipment declined, as did tankship utilization. Moreover, there was an important change in the composition of demand for oil products. These developments affected the substitutability of various crudes and significantly lowered the price differential which made one crude more attractive than another.[32]

Since price determined output shares within OPEC, small price differentials dramatically affected production in various exporting countries during the post-embargo period. That is to say, the relative price structure for OPEC crudes in 1973 was unrelated to market conditions which prevailed in 1974. Thus differences in quality, in government fiscal systems (until the end of 1974), in production costs, and in freight rates during 1974-75 produced differentials of more than $1/bbl. for crudes of relatively similar characteristics.[33] The implications of such price differentials are shown in Table 10. Table 10 shows that production during July-December 1974 increased slightly in Iran, which was attempting to increase revenues, as well as in Saudi Arabia, which was pursuing a policy of trying to moderate price increases. The greatest declines in output were in Kuwait, Libya, and the UAE. Kuwait's drop in output was the result of a decision by its government to cut back production, but the declines in production in the UAE and Libya reflected the fact

---

[31] Ian Seymour, "OPEC Adopts a New Pricing System Based on Average Government Take of $10.12/barrel for First Nine Months of 1975," Supplement to *MEES*, 13 December 1974, p. 6. Seymour's figure is based upon a production cost of 12 cents/bbl.

[32] Mabro, "OPEC after the Oil Revolution," p. 195.

[33] Mikdashi, "The OPEC Process," p. 210.

INTRA-OPEC BARGAINING AND THE PRICE OF CRUDE

Table 10

CRUDE OIL PRODUCTION BY OPEC COUNTRIES: SEPTEMBER 1973-DECEMBER 1974

Country	Production (million b/d) Average: September 1973	Production (million b/d) Average: July-December 1974	Percent Change: September 1973 to July-December 1974	Production (million b/d) Average: December 1974	Percent Change: September 1973 to December 1974
Saudi Arabia	8.57	8.63	+ 0.7%	8.05	- 6.1%
Iran	5.83	5.90	+ 1.2	5.94	+ 0.6
Kuwait	3.53	2.21	-37.4	2.32	-34.3
Iraq	2.11	1.94	- 8.0	2.18	+ 3.3
UAE	1.40	1.36	- 2.8	1.21	-13.6
Qatar	0.60	0.52	-13.3	0.52	-13.2
Libya	2.29	1.22	-46.7	1.00	-56.3
Algeria	1.10	0.97	-11.8	0.90	-18.2
Venezuela	3.39	2.82	-16.8	2.70	-20.3
Nigeria	2.14	2.24	+ 4.7	2.10	- 1.8
Indonesia	1.42	1.34	- 5.6	1.10	-22.5
Total	32.38	29.15	- 9.97	28.02	-13.46

Source: Mabro, "Can OPEC Hold the Line?," Supplement to MEES, 28 February 1975, p. 4, Table 1.

their crudes were overpriced. The quality and location premiums once commanded by these crudes fell significantly.[34]

When the UAE and Libya reduced taxes in early 1975, their output expanded rapidly. It would be wrong to characterize the actions of Libya and the UAE as a form of oligopolistic price competition: their actions were attempts to maintain output shares through price adjustments. Saudi Arabian production policy allowed considerable latitude to other countries in making price adjustments; Saudi production figures show that by mid-1975 it permitted considerable fluctuations in its own production in order to maintain OPEC prices.[35]

D. CONCLUSION

In 1973-75, the Saudi Arabian government dominated producer government bargaining on the price of crude oil. Saudi Arabian objectives shifted from conciliation with the United States to the maintenance of OPEC price discipline. Despite pressure from the United States on the Saudi government to expand production, Saudi Arabia designed its production policy to maintain the higher levels of government revenues that had been agreed on by OPEC, and thereby serve the objective of maintaining its strong position in the Arab world. During the embargo, the Saudi Arabian government demonstrated that it would risk a confrontation with the United States if the alternative was a loss of influence and prestige in the Arab world; post-embargo Saudi production policy followed the same pattern. The Saudi government—even with its low production costs and vast reserves—concluded that it was to its advantage to support OPEC; accordingly, it made sacrifices in terms of current (and potential future) revenues in order not to destroy OPEC.

In the final analysis, although the power of Saudi Arabia to influence the price of crude stemmed from economic factors—namely, Saudi Arabia's large productive capacity and its low production costs—that power was used to further essentially political objectives. In the post-1973 period, the Saudi government took steps to maintain its dominant market position, and continued a

---

[34]Mabro, "Can OPEC Hold the Line?," Supplement to *MEES*, 28 February 1975, p. 4.

[35]By April 1975 Saudi Arabia was producing at 5.65 million b/d, or about one-half of its productive capacity (*PIW*, 12 May 1975, p. 1).

program of new drilling initiated in 1971. By mid-1975 Saudi Arabia's productive capacity was almost 12 million b/d.[36]

---

[36]Francisco R. Parra, "The Pricing of Oil in International Markets," Supplement to *MEES*, 17 October 1975, p. 5. See also Economist Intelligence Unit, *Multinational Business*, June 1974, p. 37.

GENERAL CONCLUSIONS

This study has been concerned primarily with the elements that explain the shift during the 1970-75 period in the balance of power in the market for Middle Eastern crude from the international oil companies to the oil producing countries. We draw the following conclusions:

1. In conformity with the model developed in this study, the price of Middle Eastern crude during the period reflected economic forces and political decisions by the major actors in the market.

2. The economic force that was the most significant in determining price was the increasing dependence of the Western consuming countries upon Middle Eastern sources of supply.

In 1973, for example, OPEC controlled 86 percent of world oil exports; in addition, the price elasticity of oil supply and demand was extremely low, and the concentrated and vertically integrated structure of the international petroleum industry restricted price competition. This combination of economic circumstances created a potential for the exercise of monopoly power by the exporting countries.

3. Political developments largely explain the process by which the OPEC countries organized market control, as well as the timing and degree of price increases.

The seizure by a radical military regime in Libya in 1970 of the tactical advantages presented by a short-term transportation crisis to coerce the oil companies into paying higher taxes demonstrated the strategic weaknesses of the companies and of the consuming countries. The success of the Libyan government in obtaining higher taxes encouraged other producing countries to make similar demands.

The power of the producing countries to determine price increased steadily without effective challenge by the companies or the consuming countries. On the one hand, the companies discovered that after the Teheran/Tripoli negotiations, higher producing government taxes had not diminished their margins. On the other hand, the consuming countries remained divided on oil policy, and the foreign policy of the United States inadvertently strengthened the negotiating position of the oil exporting countries.

4. The October 1973 Middle East war acted as a catalyst for full assertion by OPEC countries of control over the pricing of their oil exports.

The Arab oil embargo which followed the October war was merely the last in a series of efforts by Arab exporting countries to use their monopoly power to influence U. S. foreign policy. Although the embargo failed to achieve this political objective, Arab limitations on production created the market conditions in which OPEC acted for the first time as an effective cartel.

5. Despite the continuing diplomatic efforts of the consuming countries to lower the price of oil, the level of oil prices in the 1973-75 period was largely determined by bargaining among the producing countries.

Saudi Arabia's large productive capacity, combined with its minimum needs for revenue, endowed it with preponderant bargaining strength within OPEC. Saudi Arabia used its bargaining strength primarily to maintain its leadership role in the Arab world—a political rather than economic objective.

BIBLIOGRAPHY

This bibliography is selective, and a number of items which were consulted (such as the annual reports of the oil companies) are not listed. The bibliography is divided into (A) published sources and (B) unpublished sources, and published sources are further divided as follows: (1) bibliographies, (2) general books and articles, (3) public and international documents, and (4) periodicals and trade journals. A number of references are annotated.

A. PUBLISHED SOURCES

Bibliographies

U. S. Federal Energy Administration, International Modeling and Forecasting Division, Office of Quantitative Methods. *Selected Sources of International Statistical Data on Energy: An Annotated Bibliography.* Washington: FEA, October 1974.

U. S. Library of Congress, Congressional Research Service, Economics Division. *Special Provisions of the Federal Income Tax Affecting the Oil and Gas Industry: A Summary of Provisions, Pros and Cons, and Selected References,* by Jane Gravelle. [Publication 72-189E.] Washington: Library of Congress, 25 August 1972.

U. S. Library of Congress, Congressional Research Service, Library Services Division. "Foreign Policy Implications of the Energy Crisis: A Selected Bibliography," by William E. Towsey, Jr. In U. S. Congress, House, Committee on Foreign Affairs, Subcommittee on Foreign Economic Policy. *Foreign Policy Implications of the Energy Crisis,* Hearings, 92nd Cong., 1st sess., pp. 425-455. Washington: GPO, 1972.

General Books and Articles

Adelman, M. A. *The World Petroleum Market.* Baltimore: Johns Hopkins University Press, 1972.
  Encyclopedic examination of the economics of the international

oil industry, but with unsatisfactory discussion of political developments.

――――――. "Foreign Oil: A Political-Economic Problem," *Technology Review* 76 (March-April 1974): 42-47.

――――――. "How to Have an Oil Crisis—A One-Year-Later Critique," *The Conference Board Record* 12 (January 1975): 44-46.

――――――. "Is the Oil Shortage Real? Oil Companies as OPEC Tax Collectors," *Foreign Policy* 9 (Winter 1972-1973): 69-107. Argument that U. S. policy was largely responsible for the OPEC successes in price negotiations during 1970-72.

――――――. "Politics, Economics, and World Oil," *American Economic Review* 64 (May 1974): 58-67.

――――――. "'World Oil' and the Theory of Industrial Organization." In *Industrial Organization and Economic Development; In Honor of E. S. Mason*, eds. J. W. Markham and G. F. Papanek, pp. 136-63. Boston: Houghton Mifflin, 1970.

―――――― and Friis, Soren. "Changing Monopolies and European Oil Supplies: The Shifting Balance of Economic and Political Power in the World Oil Market," *Energy Policy* 2 (December 1974): 275-92.

Akhtarekhavari, Farid. *Die Ölpreispolitik der OPEC-Länder: Grenzen, Gründe und Hintergründe*. Deutsches Übersee-Institut, Probleme der Weltwirtschaft, Diskussionbeiträge no. 2. Munich: Weltforum Verlag, 1974.

Akins, James E. "The Oil Crisis: This Time the Wolf is Here," *Foreign Affairs* 51 (April 1973): 462-90.
A defense by a U. S. State Department official of the U. S. role in the 1970-72 price negotiations.

――――――. "Can We Depend on 'Cheap Foreign Oil'?," *The Conference Board Record* 9 (July 1972): 23-24.

Alker, Hayward R., Bloomfield, Lincoln P., and Choucri, Nazli. *Analyzing Global Interdependence*. 4 vols. Cambridge, Mass.: Center for International Studies, Massachusetts Institute of Technology, November 1974. [Nos. C/74-27; C/74-28; C/74-29; C/74-30.]

Alnasrawi, Abbas. "Collective Bargaining Power in OPEC," *Journal of World Trade Law* 7 (March-April 1973): 188-207.

Amouzegar, Jamshid. "Ölpreis und weltwirtschaftliches Gleichgewicht," *Europa Archiv* 29 (10 Mai 1974): 277-84.

GENERAL BOOKS AND ARTICLES

Arad, Uzi B.; Fedoruk, Nicholas A.; Kahn, Herman; Ruggles, Rudy L., Jr.; and Shatz, Robert H. *1973-1974 Oil Embargo: Lessons and Future Impact*. [Hudson Institute Paper HI-2035-P.] Croton-on-Hudson, N. Y.: The Hudson Institute, July 1974.

Arad, Uzi B.; Hudson, Michael; Robison, David; Shatz, Robert H.; and Smernoff, Barry J. *Energy and Security: Implications for American Policy*. [Hudson Institute Report HI-1884/2-RR.] Croton-on-Hudson, N. Y.: The Hudson Institute, July 1974.

Bergsten, C. Fred, and Krause, Lawrence B., eds. *World Politics and International Economics*. Washington, D. C.: The Brookings Institution, 1975.
Originally published as special issue of *International Organization* 29 (Winter 1975). Useful discussion of political and economic models of international economic relations.

Bohi, Douglas R., and Russell, Milton. *U. S. Energy Policy: Alternatives for Security*. Baltimore: Johns Hopkins University Press, 1975.

Brannon, Gerald M. *Energy Taxes and Subsidies*. Cambridge, Mass.: Ballinger, 1974.
Assessment of the financial implications of U. S. tax policy for Middle Eastern operations of American oil companies.

Breton, Hubert. "Le pétrole libyen au service de l'unité arabe?," *Revue Francaise de Science Politique* 22 (décembre 1972): 1256-75.

British-North American Committee. *Higher Oil Prices: Worldwide Financial Implications*. Washington, D. C.: British-North American Committee, October 1975.
Contains a research report on the size and disposition of OPEC oil revenues.

Campbell, John C. "Middle East Oil: American Policy and Super-Power Interaction," *Survival* 15 (September-October 1973): 210-18.

Chandler, Geoffrey. *Oil—Prices and Profits*. [Discussion Paper No. 13.] London: Foundation for Business Responsibilities, 1975.
Statistical information regarding producing government revenues and company margins.

―――――――. "The Myth of Oil Power: International Groups and National Sovereignty," *International Affairs* 46 (October 1970): 710-18.
Suggests reasons for the declining bargaining power of the majors.

BIBLIOGRAPHY

Chenery, Hollis B. "Restructuring the World Economy," *Foreign Affairs* 53 (January 1975): 242-63.

Chevalier, Jean-Marie. *The New Oil Stakes*. Trans. by Ian Rock. London: Allen Lane, 1975. [Published in 1973 by Calmann-Lévy as *Le nouvel enjeu pétrolier*.]
Unconvincing argument that "international financial capitalism" will monopolize alternative energy sources.

Committee for Economic Development, Research and Policy Committee. *International Economic Consequences of High-Priced Energy*. New York: Committee for Economic Development, 1975.

Connery, Robert H., and Gilmour, Robert S., eds. "The National Energy Problem," *Proceedings of the Academy of Political Science* 31 (December 1973): 1-194.

Cooper, Richard N., ed. *A Reordered World: Emerging International Economic Problems*. Washington, D.C.: Potomac Associates, 1973.
Includes Adelman, "Is the Oil Shortage Real?," and Levy, "An Atlantic-Japanese Energy Policy."

Cooperative Approaches to World Energy Problems: A Tripartite Report by Fifteen Experts from the European Community, Japan and North America. Washington, D. C.: The Brookings Institution, 1974.
Summary of the basic issues involved in consuming country cooperation on oil prices.

Corden, W. M. "Implications of the Oil Price Rise," *Journal of World Trade Law* 8 (March-April 1974): 133-43.

Darmstadter, Joel, and Landsberg, Hans H. "The Economic Background," *Daedalus* 104 (Fall 1975): 15-38.

De Carmoy, Guy. *The Oil Crisis and the Energy Policy of the Industrial Nations*. [Research Papers, European Series No. E104.] Fontainebleau: European Institute of Business Administration, June 1973.

Demetz, Harold. *The Market Concentration Doctrine: An Examination of Evidence and a Discussion of Policy*. [Hoover Policy Study No. 7.] Washington, D. C.: American Enterprise Institute, 1973.

H. P. Drewry (Shipping Consultants). *Host Government Participation in the Oil Trade*. [No. 20 in a Series.] London: H. P. Drewry, 1974.
Statistics on direct crude sales by OPEC countries.

Duchesneau, Thomas D. *Competition in the U. S. Energy Industry*.

Cambridge, Mass.: Ballinger, 1975.
Demonstrates that government-imposed restrictions were responsible for the rapid growth in U. S. oil imports.

Ebel, Robert E. *Communist Trade in Oil and Gas: An Evaluation of the Future Export Capability of the Soviet Bloc.* New York: Praeger, 1970.

Elm, Mostafa. "Oil Negotiations: A View from Iran," *Columbia Journal of World Business* 6 (November-December 1971): 81-90.

Erickson, Edward W., and Waverman, Leonard, eds. *The Energy Question: An International Failure of Policy.* 2 vols. Toronto: University of Toronto Press, 1974.
R. Mabro, "Political and Financial Aspects of the Oil Game," assesses the embargo, and Maureen S. Crandall, "Oil in the Middle East and North Africa," surveys developments during 1960-74.

Energy Policy Project of the Ford Foundation. *A Time to Choose: America's Energy Future.* Cambridge, Mass.: Ballinger, 1974.
Superficial treatment of the price and supply of Middle Eastern crude.

Exxon Company. *Competition in the Petroleum Industry.* Houston: Exxon Company, 1975.
Reprint of the testimony of W. T. Slick, Senior Vice President, Exxon Company, before the Senate Judiciary Subcommittee on Antitrust and Monopoly, 21 January 1975.

Field, Michael. "Oil in the Middle East and North Africa." In *The Middle East and North Africa 1975-1976.* 22d ed. London: Europa Publications, 1975.

_____. "Oil: OPEC and Participation," *The World Today* 28 (January 1972): 5-13.

First, Ruth. *Libya: The Elusive Revolution.* Harmondsworth: Penguin Books, 1974.
Analysis of the political motives underlying the oil policies of the RCC.

Frank, Helmut J. *Crude Oil Prices in the Middle East: A Study in Oligopolistic Price Behavior.* New York: Praeger, 1966.

Frankel, Paul H. *Essentials of Petroleum: A Key to Oil Economics.* London: Frank Cass, 1976.
Basic statement of the principles upon which the oil industry is organized.

BIBLIOGRAPHY

Fritsch, Albert J., and Egan, John W. *Big Oil: A Citizen's Factbook on the Major Oil Companies.* Washington, D. C.: Center for Science in the Public Interest, 1973.
Compilation of data regarding sales, profits, and assets of the American majors.

Fried, Edward R., and Schultze, Charles L., eds. *Higher Oil Prices and the World Economy: The Adjustment Problem.* Washington, D. C.: The Brookings Institution, 1975.

Friedland, Edward; Seabury, Paul; and Wildavsky, Aaron. "Oil and the Decline of Western Power," *Political Science Quarterly* 90 (Fall 1975): 437-50.

Friedrich-Ebert-Stiftung. *Zukunftsorientierte Energie- und Rohstoffpolitik.* Bonn-Bad Godesberg: Verlag Neue Gesellschaft, 1976. [Proceedings of a conference of the Freidrich-Ebert-Stiftung, 13-14 October 1975, Bonn.]
Paul H. Frankel, "Organisationsprobleme der Ölimportländer," discusses OPEC problems in organizing market control.

Gordon, Richard L. *The Evolution of Energy Policy in Western Europe: The Reluctant Retreat from Coal.* New York: Praeger, 1970.

Graubard, Stephen R., ed. "The Oil Crisis: In Perspective," *Daedalus* 104 (Fall 1975): 1-291.
Collection of sixteen essays that constitute a provocative and well-documented analysis of the causes and implications of the 1973-74 embargo.

Hansen, Joachim. "Neue Ära der Internationalen Ölwirtschaft," *Aussenpolitik* 24 (2 Quartal 1973): 201-10.

_____. "Die Herausforderung der OPEC-Länder," *Aussenpolitik* 22 (März 1971): 129-239.

Hartshorn, J. E. *Oil Companies and Governments: An Account of the International Industry in Its Political Environment.* London: Faber and Faber, 1962.
Clear and comprehensive statement of the nature of bargaining over price which identifies the divergent interests of the major actors in the market.

_____. "Oil Diplomacy: The New Approach," *The World Today* 29 (July 1973): 281-90.

_____. "From Tripoli to Teheran and Back: The Size and Meaning of the Oil Game," *The World Today* 27 (July 1971): 291-

GENERAL BOOKS AND ARTICLES

301.

Analysis of the changing nature of consumer-producer bargaining and of the role of spare capacity in determining negotiating strength.

Hill, Peter, and Vielvoye, Roger. *Energy in Crisis: A Guide to World Oil Supply and Demand and Alternative Sources.* London: Robert Yeatman, 1974.

Compilation of data concerning the economic impact of the embargo and of higher oil prices, with particular reference to the tanker market.

Hirschman, Albert O. *National Power and the Structure of Foreign Trade.* Berkeley: University of California Press, 1975.

The first chapter reviews the theory of the relationship between power and international trade.

International Institute for Strategic Studies (IISS). *The Middle East and the International System: II. Security and the Energy Crisis.* [Adelphi Paper No. 115.] London: IISS, 1975.

_____. *The Middle East and the International System: I. The Impact of the 1973 War.* [Adelphi Paper No. 114.] London: IISS, 1975.

Iskandar, Marwan. *The Arab Oil Question.* Beirut: Aleph, 1973.

Issawi, Charles. *Oil, the Middle East and the World.* [Washington Papers, Vol. 1, No. 4.] Beverly Hills: Sage Publications, 1972.

Itayim, Fuad. "Strengths and Weaknesses of the Oil Weapon." In IISS, *The Middle East and the International System: II. Security and the Energy Crisis.* [Adelphi Paper No. 115.] London: International Institute for Strategic Studies, 1975.

Contains an assessment of the embargo that differs from this study.

Jacoby, Neil E. *Multinational Oil: A Study in Industrial Dynamics.* New York: Macmillan, 1974.

An appraisal of market structure and performance in the international petroleum industry that demonstrates a significant decrease in market concentration during 1960-73.

Kahn, Herman. *Oil Prices and Energy in General.* [Hudson Institute Paper HI-2063/3-P.] Croton-on-Hudson, N. Y.: The Hudson Institute, August 1974.

Summarizes U. S. and OECD policy failures leading to the 1973-74 embargo.

BIBLIOGRAPHY

Kenen, Peter B. "Oil, Inflation, and Economic Policy," *Near East Report* 18 (16 October 1974): 213-16.

Kennedy, Edward M. "The Persian Gulf: Arms Race or Arms Control?," *Foreign Affairs* 54 (October 1975): 14-35.

Klebanoff, Shoshana. *Middle East Oil and U. S. Foreign Policy: With Special Reference to the U. S. Energy Crisis.* New York: Praeger, 1974.
History of U. S. oil policy since 1930.

Knorr, Klaus. "The Limits of Economic and Military Power," *Daedalus* 104 (Fall 1975): 229-44.

Krasner, Stephen D. "Oil Is the Exception," *Foreign Policy* 14 (Spring 1974): 63-83.

Krause, Lawrence B., and Nye, Joseph S. "Reflections on the Economics and Politics of International Economic Organizations." In *World Politics and International Economics*, eds. C. F. Bergsten and L. B. Krause, pp. 323-42. Washington, D. C.: The Brookings Institution, 1975.

Krueger, Robert B. *The United States and International Oil: A Report to the Federal Energy Administration on U. S. Firms and Government Policy.* New York: Praeger, 1975.
Part II contains an assessment of post-1970 U. S. policy toward Middle Eastern producers.

Kubbah, Abdul Amir Q. *OPEC: Past and Present.* Vienna: Petro-Economic Research Centre, [1974].
Analysis of the role of OPEC in bargaining over price.

Laqueur, Walter. *Confrontation: The Middle East and World Politics.* New York: Bantam Books, 1974.
Lucid account of the 1973 war.

_____. *The Struggle for the Middle East: The Soviet Union and the Middle East 1958-1970.* Harmondsworth: Penguin Books, 1969.

Lenczowski, George. *Oil and the State in the Middle East.* Ithaca, N. Y.: Cornell University Press, 1960.
Contains an analysis of the economic and political implications of the 1957 Suez crisis.

_____. "Multinational Oil Companies: A Factor in Middle East International Relations," *California Management Review* 8 (Winter 1970): 38-44.

Suggests that control of transportation and of alternative sources of supply are the controlling factors in company negotiating strength.

───────────. "The Oil-Producing Countries," *Daedalus* 104 (Fall 1975): 59-72.

Levy, Walter J. "An Atlantic-Japanese Energy Policy," *Foreign Policy* 11 (Summer 1973): 159-90.
Increased consuming country cooperation is deemed necessary to meet the growing bargaining power of the producing countries.

───────────. "Oil Power," *Foreign Affairs* 49 (July 1971): 652-68.

Lubell, Harold. *Middle East Oil Crises and Western Europe's Energy Supplies*. Baltimore: Johns Hopkins, 1963.
Part I reviews supply interruptions in the 1950's.

Lufti, Ashraf. *OPEC Oil*. Beirut: The Middle East Research and Publishing Center, 1968.

Mabro, Robert E. "Libya." In *A Survey of North West Africa*, ed. W. Knapp. Oxford: Oxford University Press, 1976.

───────────. "OPEC After the Oil Revolution," *Millennium* 4 (Winter 1975-1976): 191-99.

Madelin, Henri. *Oil and Politics*. Trans. by Margaret Totman. Lexington, Mass.: D. C. Heath, 1975. [Published in 1973 by Armand Colin as *Pétrole et politique en Mediterranée Occidentale*.]
Study of the divergent interests among exporting and consuming governments, with special emphasis on Libya and Algeria.

Maffre, John. "Economic Report/Administration Searching for Methods to Aid U. S. Oil Interests in Bargaining with OPEC," *National Journal* 4 (13 May 1972): 808-18.
A good summary of U. S. policies during the 1970-71 price negotiations.

Mancke, Richard B. *The Failure of U. S. Energy Policy*. New York: Columbia University Press, 1974.
Provides a brief treatment of the basic principles of U. S. tax and oil-import policies.

───────────. *Performance of the Federal Energy Office*. [National Energy Study No. 6.] Washington, D. C.: American Enterprise Institute, 1975.

───────────. "The Future of OPEC," *The Journal of Business* 48 (January 1975): 11-19.

BIBLIOGRAPHY

Manoharan, S. *The Oil Crisis: End of an Era.* New Delhi: S. Chand, 1974.

Maull, Hanns. *Oil and Influence: The Oil Weapon Examined.* [Adelphi Paper No. 117.] London: International Institute for Strategic Studies, 1975.
Espouses the view that the embargo of 1973-74 caused a fundamental change in the international political system.

McKie, James W. "The Political Economy of World Petroleum," *American Economic Review* 64 (May 1974): 51-57.

_____. "The United States," *Daedalus* 104 (Fall 1975): 73-90.

Menderhausen, Horst, and Nehring, Richard. *Protecting the U. S. Petroleum Market Against Future Denials of Imports.* [Rand Report R-1603-ARPA.] Santa Monica: The Rand Corporation, October 1974.
Assessment of the U. S. policy response to the embargo.

The Middle East and North Africa 1975-1976. 22d ed. London: Europa Publications, 1975.

Mikdashi, Zuhayr. *The Community of Oil-Exporting Countries: A Study in Governmental Cooperation.* London: George Allen and Unwin, 1972.
Study of objectives, structure, and performance of OPEC since its establishment.

_____. *A Financial Analysis of Middle Eastern Oil Concessions: 1901-1965.* New York: Praeger, 1966.

_____. "Cooperation Among Oil Exporting Countries with Special Reference to Arab Countries: A Political Economy Analysis," *International Organization* 28 (Winter 1974): 1-30.

_____. "The OPEC Process," *Daedalus* 104 (Fall 1975): 203-16.

_____; Cleland, Sherril; and Seymour, Ian, eds. *Continuity and Change in the World Oil Industry.* Beirut: Middle East Research and Publishing Center, 1970.
A collection of papers presented at the Third Seminar on the Economics of the International Petroleum Industry, the American University of Beirut, Spring 1969.

Mitchell, Edward J. *U. S. Energy Policy: A Primer.* [National Energy Study No. 1.] Washington, D. C.: American Enterprise Institute,

GENERAL BOOKS AND ARTICLES

1974.
Useful assessment of post-1973 U. S. import policies.

──────────, ed. *Dialogue on World Oil: A Conference Sponsored by the American Enterprise Institute's National Energy Project.* Washington, D. C.: American Enterprise Institute, 1974.

Monroe, Elizabeth, and Farrar-Hockley, A. H. *The Arab-Israel War, October 1973: Background and Events.* [Adelphi Paper No. 111.] London: International Institute for Strategic Studies, 1975.

────────── and Mabro, Robert. *Oil Producers and Consumers: Conflict or Cooperation: An International Seminar Report.* New York: American Universities Field Staff, 1974. [Synthesis of an international seminar at the Center for Mediterranean Studies, Rome, 24-28 June 1974.]

Mosley, Leonard. *Power Play: Oil in the Middle East.* Baltimore: Penguin Books, 1973.
Journalistic account of oil and politics in the Middle East.

National Petroleum Council, Committee on Energy Conservation. *Potential for Energy Conservation in the United States: A Summary Report.* [Washington, D. C.]: National Petroleum Council, 10 September 1974.

No Time to Confuse: A Critique of the Final Report of the Energy Policy Project of the Ford Foundation: "A Time to Choose: America's Energy Future." San Francisco: Institute for Contemporary Studies, 1975.
Ten essays dealing with fundamental analytical errors in the Report of the Energy Policy Project.

Nordhaus, William O. "The Allocation of Energy Resources," *Brookings Papers on Economic Activity*, No. 3 (1973), pp. 529-70.

Odell, Peter R. *The Western European Energy Economy: Challenges and Opportunities.* London: University of London, Athlone Press, 1975.

──────────. *Oil and World Power: Background to the Oil Crisis.* 3d ed. Harmondsworth: Penguin Books, 1974.
Provocative discussion of the post-1973 strategies of the consuming and producing governments.

Otaiba, Mana Saeed Al. *OPEC and the Petroleum Industry.* London: Croom Helm, 1975.
The author, the current minister of petroleum of the UAE, identifies OPEC's chief weakness as its inability to devise a producing

sharing program.

Pen, J. "Bilateral Monopoly, Bargaining and the Concept of Economic Power." In *Power in Economics: Selected Readings*, ed. K. W. Rothschild. Harmondsworth: Penguin Books, 1971.

Penrose, Edith T. *The Growth of Firms, Middle East Oil and Other Essays*. London: Frank Cass, 1971.

_____. *The International Oil Industry in the Middle East*. Cairo: National Bank of Egypt, 1968.
Useful analysis of the role of OPEC and of its effect upon the structure of the international petroleum industry.

_____. *The Large International Firm in Developing Countries: The International Petroleum Industry.* London: George Allen and Unwin, 1968.
Employs an institutional approach and carefully documents the erosion of market control exercised by the majors.

_____. "The Development of Crisis," *Daedalus* 104 (Fall 1975): 39-58.

_____. "Origins and Development of the International Oil 'Crisis'," *Millennium* 3 (January 1974): 37-43.

Petroleum Information Foundation, Inc. *Background Information Paper No. 16*. New York: Petroleum Information Foundation, October 1973.
Compilation of information on producing government oil revenues 1960-72.

_____. *Oil Imports and the National Interest*. New York: Petroleum Information Foundation, March 1971.

Prodi, Romano, and Clô, Alberto. "Europe," *Daedalus* 104 (Fall 1975): 91-112.

Ray, George F. *Western Europe and the Energy Crisis*. [Thames Essay No. 6.] London: Trade Policy Research Centre, January 1975.

Rifaï, Taki. *The Pricing of Crude Oil: Economic and Strategic Guidelines for an International Energy Policy*. New York: Praeger, 1974.
The author was an adviser to the Libyan government, and offers some insight into its bargaining strategies in the price negotiations.

_____. "La crise pétrolière internationale (1970-1971): Essai d'interprétation," *Revue Francaise de Science Politique* 22 (décembre 1972): 1205-36.

Robinson, Colin. *The Energy "Crisis" and British Coal.* [Hobart Paper No. 59.] London: Institute of Economic Affairs, 1974.
    Maintains that the internal conditions which usually cause the breakup of cartels are absent in OPEC.

Roeber, Richard J. C. *The Organization in a Changing Environment.* Reading, Mass.: Addison-Wesley, 1973.
    Treatment of the oil industry as a case study of an organization which failed to adapt to a changing political environment.

Rouhani, Fuad. *A History of O.P.E.C.* New York: Praeger, 1971.

Ruttenberg, Stanley H., and Associates. *The American Oil Industry: A Failure of Anti-Trust Policy.* New York: Marine Engineers' Beneficial Association, December 1973.
    Contains useful information on joint ventures among major oil companies.

Safer, Arnold E., and Mills, Anne Parker. *Outlook for World Oil: Prices and Petrodollars.* New York: Irving Trust Company, 20 March 1975.

Sampson, Anthony. *The Seven Sisters: The Great Oil Companies and the World They Made.* London: Hodder and Stoughton, 1975.
    Readable account of 1970-75 developments, based in part upon the hearings of the Senate Subcommittee on Multinational Corporations.

Scherer, F. M. *Industrial Market Structure and Economic Performance.* Chicago: Rand McNally, 1970.

Schuler, George Henry Mayer. "The International Oil 'Debacle' Since 1971," Special Supplement to *Petroleum Intelligence Weekly*, 22 April 1974, pp. 1-36.
    Reprint of the testimony of the LPG representative from Bunker Hunt before the Senate Subcommittee on Multinational Corporations.

Schurr, Sam H., and Homan, Paul T. *Middle Eastern Oil and the Western World: Prospects and Problems.* New York: American Elsevier, 1971.
    Detailed review of economic developments and company-government relations during 1960-69.

Shwadran, Benjamin. *The Middle East, Oil and the Great Powers.* 3d ed. New York: John Wiley, 1973.

Simonet, Henri. "Energy and the Future of Europe," *Foreign Af-

fairs 53 (April 1975): 450-63.

Stobaugh, Robert B. "The Oil Companies in the Crisis," *Daedalus* 104 (Fall 1975): 179-202.

Stocking, George W. *Middle East Oil: A Study in Political and Economic Controversy*. London: Allen Lane, 1970.
A conventional treatment of price negotiations since the discovery of oil in the Middle East.

Stork, Joe. *Middle East Oil and the Energy Crisis*. New York: Monthly Review Press, 1971.
Neo-Marxist analysis of developments in the Middle East 1960-74, with detailed description of Iraqi policies.

Szyliowicz, Joseph S. "The Embargo and U. S. Foreign Policy." In *The Energy Crisis and U. S. Foreign Policy*, eds. J. S. Szyliowicz and B. E. O'Neill. New York: Praeger, 1975.

_____ and O'Neill, Bard E. *The Energy Crisis and U. S. Foreign Policy*. New York: Praeger, 1975.
Collection of ten essays of uneven quality assessing U. S. policy during the 1973-74 embargo.

Tahtinen, Dale R. *Arms in the Persian Gulf*. [Foreign Affairs Studies No. 10.] Washington, D. C.: American Enterprise Institute, 1974.

Tsurumi, Yoshi. "Japan," *Daedalus* 104 (Fall 1975): 113-28.

Tugendhat, Christopher, and Hamilton, Adrian. *Oil: The Biggest Business*. Rev. ed. London: Eyre Methuen, 1975.
A provocative treatment of the shifting balance of power in the world oil market and its implications for company profitability.

Tugwell, Franklin. *The Politics of Oil in Venezuela*. Stanford: Stanford University Press, 1975.
Discussion of "interdependent bargaining" between oil companies and governments.

von Geusau, Frans A. M. Alting, ed. *Energy in the European Communities*. Leyden: A. W. Sijhoff, 1975.
Identifies the problems of cooperation among Western European consuming governments.

Wasserman, Ursula. "The Energy Crisis," *Journal of World Trade Law* 8 (September-October 1974): 364-87.

Wilkins, Mira. "The Oil Companies in Perspective," *Daedalus* 104 (Fall 1975): 159-78.

Windsor, Philip. *Oil: A Plain Man's Guide to the World's Energy*

Crisis. London: Maurice Temple Smith, 1975.

Yager, Joseph A., and Steinberg, Eleanor B. *Energy and U. S. Foreign Policy.* Cambridge, Mass.: Ballinger, 1974.
Includes an assessment of the objectives and strategies of the Arab exporters during 1970-74.

Yamani, Sheikh Ahmed Zaki. "Oil: Towards a New Producer-Consumer Relationship," *The World Today* 30 (November 1974): 479-86.

───────────. "Participation versus Nationalization." In *Continuity and Change in the World Oil Industry*, eds. Z. Mikdashi, S. Cleland, and I. Seymour. Beirut: Middle East Research and Publishing Center, 1970.

Public and International Documents

Iran, Ministry of Information. *Text of Interviews Granted by H.I.M. the Shahanshah Aryamehr.* Teheran: Publications Department, Ministry of Information. September 1974.

MNC Hearings. See U. S. Congress, Senate, Committee on Foreign Relations, Subcommittee on Multinational Corporations.

Organization for Economic Cooperation and Development (OECD). *Energy Prospects to 1985: An Assessment of Long-Term Energy Developments and Related Policies.* 2 vols. Paris: OECD, 1974.

───────────. *OECD Economic Outlook.* Biannual.

───────────, Oil Committee. *Oil Statistics: Supply and Disposal.* Annual.

United Nations, Department of Economic and Social Affairs, Ad Hoc Panel of Experts on Projections of Demand and Supply of Crude Petroleum and Products. *Petroleum in the 1970's* (ST/ECA/179) 1974.

U. S. Cabinet Task Force on Oil Import Control. *The Oil Import Question: A Report on the Relationship of Oil Imports to the National Security.* Washington: GPO, 1970.

U. S. Comptroller General. *Issues Related to Foreign Sources of Oil for the United States: Department of State.* Report to the Congress. Washington: U. S. General Accounting Office, 23 January 1974. [GAO Report B-179411]

U. S. Congress, House, Committee on Banking, Currency and Hous-

BIBLIOGRAPHY

ing, Ad Hoc Committee on the Domestic and International Monetary Effect of Energy and Other Natural Resource Pricing. *The Economics of Energy and Natural Resource Pricing.* Compilation of Reports and Hearings, 94th Cong., 1st sess. Washington: GPO, 1975. [Committee print]

_____. *Meeting America's Resource Needs: Problems and Policies.* Hearings, 93d Cong., 2d sess. Washington: GPO, 1974.

_____. *Oil Imports and Energy Security.* Hearings, 93d Cong., 2d sess. Washington: GPO, 1974.

U. S. Congress, House, Committee on Foreign Affairs. *The United States Oil Shortage and the Arab-Israeli Conflict.* Report of a Study Mission to the Middle East from October 22 to November 3, 1973, 93d Cong., 1st sess. Washington: GPO, 1973. [Committee print]
Contains a useful chronology of events during the early months of the embargo.

U. S. Congress, House, Committee on Foreign Affairs, Subcommittees on Europe and on the Near East and South Asia. *United States-Europe Relations and the 1973 Middle East War.* Hearings, 93d Cong., 1st and 2d sess. Washington: GPO, 1974.

U. S. Congress, House, Committee on Foreign Affairs, Subcommittee on Foreign Economic Policy. *Foreign Policy Implications of the Energy Crisis.* Hearings, 92d Cong., 1st sess. Washington: GPO, 1972.

U. S. Congress, House, Committee on Foreign Affairs, Subcommittees on Foreign Economic Policy and on the Near East and South Asia. *Oil Negotiations, OPEC, and the Stability of Supply.* Hearings, 93d Cong., 1st sess. Washington: GPO, 1973.

U. S. Congress, House, Committee on Foreign Affairs, Subcommittees on International Organizations and Movements and on Foreign Economic Policy. *U. S. Policy and the International Energy Agency.* Hearings, 93d Cong., 2d sess. Washington: GPO, 1975.

U. S. Congress, House, Committee on Foreign Affairs, Subcommittee on the Near East. *U. S. Interests in and Policy Toward the Persian Gulf.* Hearings, 92d Cong., 2d sess. Washington: GPO, 1972.

U. S. Congress, House, Committee on Foreign Affairs, Subcommittee on the Near East and South Asia. *The Persian Gulf, 1974: Money, Politics, Arms and Power.* Hearings, 93d Cong., 2d sess. Washington: GPO, 1975.

PUBLIC AND INTERNATIONAL DOCUMENTS

_____. *The Middle East, 1974: New Hopes, New Challenges.* Hearings, 93d Cong., 2d sess. Washington: GPO, 1975.

_____. *New Perspectives on the Persian Gulf.* Hearings, 93d Cong., 2d sess. Washington: GPO, 1973.

U. S. Congress, House, Committee on International Relations, Subcommittees on International Organizations and on International Resources, Food and Energy. *Legislation on the International Energy Agency.* Hearings, 94th Cong., 1st sess. Washington: GPO, 1975.

U. S. Congress, House, Committee on Ways and Means. *The Energy Crisis and Proposed Solutions.* Panel Discussions, 94th Cong., 1st sess. Washington: GPO, 1975. 4 parts.

_____. *"Windfall" or Excess Profits Tax.* Hearings, 93d Cong., 2d sess. Washington: GPO, 1974.
A review of U. S. tax policy and the implications of higher prices for company profitability.

_____. *General Tax Reform (Testimony from Administration and Public Witnesses).* Hearings, 93d Cong., 1st sess. Washington: GPO, 1973. 18 parts.

U. S. Congress, Joint Committee on Atomic Energy. *Understanding the "National Energy Dilemma."* Report by the Staff of the Joint Committee on Atomic Energy, 93d Cong., 1st sess. Washington: GPO, 1973. [Joint Committee print]

U. S. Congress, Joint Economic Committee, Subcommittee on International Economics. *Economic Impact of Petroleum Shortages.* Hearings, 93d Cong., 1st sess. Washington: GPO, 1974.

U. S. Congress, Senate, Committee on Finance. *Energy Windfall Profits.* Hearings, 93d Cong., 2d sess. Washington: GPO, 1974.

_____. *World Oil Developments and U. S. Oil Import Policies.* A Report Prepared by the U. S. Tariff Commission, 93d Cong., 1st sess. Washington: GPO, 1973. [Committee print]
Comprehensive review of trends in market prices and U. S. import restrictions since 1957.

U. S. Congress, Senate, Committee on Foreign Relations. *Energy and Foreign Policy.* Hearings, 93d Cong., 1st sess. Washington: GPO, 1974.

U. S. Congress, Senate, Committee on Foreign Relations, Subcommittee on Multinational Corporations. *Multinational Corporations and United States Foreign Policy*. Hearings 93d Cong., 1st and 2d sess. Washington: GPO, 1975. 10 parts.

U. S. Congress, Senate, Committee on Government Operations, Permanent Subcommittee on Investigations. *Current Energy Shortages Oversight Series*. Hearings, 93d Cong., 2d sess. Washington: GPO, 1974. 8 parts.

U. S. Congress, Senate, Committee on Interior and Insular Affairs. *Energy Policy Papers*. 93d Cong., 2d sess. Washington: GPO, 1974. [Committee print]

_____. *Measurement of Corporate Profits*, by Julius W. Allen. Report, 93d Cong., 2d sess. Washington: GPO, 1974. [Committee print]
Analysis of accounting conventions used in the oil industry.

_____. *Implications of Recent Organization of Petroleum Exporting Countries (OPEC) Oil Price Increases*. A National Fuels and Energy Policy Study, 93d Cong., 1st sess. Washington: GPO, 1974. [Committee print]

_____. *Market Performance and Competition in the Petroleum Industry*. Hearings, 93d Cong., 1st sess. Washington: GPO, 1974. 2 parts.

_____. *Oil and Gas Imports Issues*. Hearings, 93d Cong., 1st sess. Washington: GPO, 1973. 3 parts.

_____. *Towards a Rational Policy for Oil and Gas Imports*. A Policy Background Paper, 92d Cong., 2d sess. Washington: GPO, 1973. [Committee print]

_____. *Considerations in the Formulation of National Energy Policy*. A National Fuels and Energy Policy Study, 92d Cong., 1st sess. Washington: GPO, 1971. [Committee print]

U. S. Department of the Interior, Bureau of Mines. *International Petroleum Annual*.

U. S. Department of the Interior, Office of Oil and Gas. *1972-1973 Estimated International Flow of Petroleum and Tanker Utilization*. Washington: Department of the Interior, May 1973.

_____. *Middle East Petroleum Emergency of 1967*. Washington: GPO, 1969. 2 parts.

_____. *Worldwide Crude Oil Prices*. Annual.

U. S. Department of the Treasury. *Staff Analysis of the Preliminary Federal Trade Commission Staff Report on its Investigation of*

PUBLIC AND INTERNATIONAL DOCUMENTS

the Petroleum Industry, July 2, 1973. Prepared for the Committee on Interior and Insular Affairs, U. S. Senate, 93d Cong., 1st sess. Washington: GPO, 1973. [Committee print]

U. S. Federal Energy Administration, National Energy Information Center. *Monthly Energy Review*.

──────────. *1972 Petroleum Supply and Demand in the Non-Communist World*. Washington: GPO, May 1974.

U. S. Federal Energy Administration, Office of International Energy Affairs. *The Relationship of Oil Companies and Foreign Governments*. Washington: GPO, June 1975.
A comprehensive assessment of the status of participation in the Middle East.

──────────. *U. S. Oil Companies and the Arab Oil Embargo: The International Allocation of Constricted Supplies*. Prepared for the Subcommittee on Multinational Corporations, Committee on Foreign Relations, U. S. Senate, 94th Cong., 1st sess. Washington: GPO, 27 January 1975. [Committee print]

U. S. Federal Energy Administration, Office of Policy and Analysis, Petroleum Industry Monitoring System (PIMS). *U. S.—OPEC Petroleum Report*. Washington: Federal Energy Administration, 1 July 1974.
Statistical information on U. S. imports from OPEC.

U. S. Federal Energy Office, Office of International Energy Affairs. *A Discussion of the World Energy Market in 1980 and 1985*. Washington: Federal Energy Office, April 1974.

U. S. Federal Energy Office, Petroleum Industry Monitoring System. *The PIMS Monthly Report*.

U. S. Library of Congress, Congressional Research Service, Foreign Affairs Division. *Chronology of the Libyan Oil Negotiations, 1970-1971*. Prepared for the Subcommittee on Multinational Corporations, Committee on Foreign Relations, U. S. Senate, 93d Cong., 2d sess. Washington: GPO, 1974. [Committee print]

U. S. Library of Congress, Congressional Research Service, Science Policy Research Division. *Energy Facts*. Prepared for the Subcommittee on Energy, Committee on Science and Astronautics, U. S. House. Washington: GPO, 1974. [Committee print]
Includes comprehensive statistics on oil and other sources of energy for 1972.

──────────. *Energy Policy and Resource Management*. Report prepared for the Subcommittee on Energy, Committee on Science and Astronautics, U. S. House, 93d Cong., 2d sess. Washington: GPO, 1974. [Committee print]

Periodicals and Trade Journals

American Academic Association for Peace in the Middle East (Washington, D. C.). *Middle East Issues Series.*

———. *Middle East Review* (formerly *Middle East Information Series*).

Arab Oil and Gas (Beirut).

British Petroleum Company (London). *BP Statistical Review of the World Oil Industry.* Annual.

Chase Manhattan Bank (New York), Energy Economics Division. *Capital Investments of the World Petroleum Industry.* Annual.

———. *Financial Analysis of a Group of Petroleum Companies.* Annual.

———. *The Petroleum Situation.* Monthly.

Daedalus. Quarterly. *See* Graubard, ed.

The Economist (London).

The Economist Intelligence Unit (London). *Multinational Business.* Quarterly.

First National City Bank (New York), Petroleum Department. *Energy Memo.* Quarterly.

W. Greenwell & Co. (London). *Oil Commentary.* Published irregularly.

Independent Petroleum Association of America (Washington, D. C.). *United States Petroleum Statistics.* Annual.

Middle East Economic Survey (MEES) (Beirut).

Oil and Gas Journal (Tulsa, Oklahoma).

Organization of Petroleum Exporting Countries (OPEC). *Annual Review and Record* (Vienna).

Parra, Ramos and Parra, S. A. *International Crude Oil and Product Prices.* Biannual.

This series, prepared in cooperation with *MEES*, includes data relating to market prices for Middle Eastern crudes from 1969.

UNPUBLISHED SOURCES

Petroleum Economist (PE) (London).
Petroleum Intelligence Weekly (PIW) (New York).
Petroleum Press Service (PPS). Monthly.

B. UNPUBLISHED SOURCES

Chandler, Geoffrey. "The Role of the International Oil Companies." An address at the Second International Colloquium on the Petroleum Economy, Laval University, Quebec, 3 October 1975. [Mimeograph]

Colitti, Marcello. "Vertical Integration, Major Oil Companies and Newcomers: The Case of ENI." Paper presented to the Oil Seminar, St. Antony's College, Oxford, 2 March 1976. [Typewritten]

Conant, Melvin A. "Convergence and Divergence between the Importing Countries: American Policies and Their Impact in Europe and Japan." An address at the Second International Colloquium on the Petroleum Economy, Laval University, Quebec, 3 October 1975. [Mimeograph]

Patton, George T. "Current Outlook for the Energy Sector." Washington, D. C.: American Petroleum Institute, October 1974. [Typewritten]

Petroleum Industry Research Foundation, Inc. "U. S. Oil Imports and Import Dependency." New York: Petroleum Industry Research Foundation, 19 December 1974. [Mimeograph]

Quadrangular Conference II; The Interrelationship Between Inflation/Recession, The International Financial Structure, and Alliance Security. "Panelists' Remarks." Organized by the Center for Strategic and International Studies, Georgetown University, Washington, D. C., 27-29 January 1975. [Mimeograph]

Second World Energy Supplies Conference. "Speakers' Papers." Organized by the *Financial Times* and the *Oil Daily*, Americana Hotel, New York, 24-26 April 1974. [Mimeograph]

U. S. Customs Service, Office of Operations, Duty Assessment Division. "Information Concerning the Pricing of International Oil." Washington: Customs Service, February 1975. [Mimeograph]

U. S. Federal Energy Administration, National Energy Information Center. "Voluntary Agreement and Program Relating to the International Energy Program." Washington: Federal Energy Admini-

BIBLIOGRAPHY

stration, 6 March 1975. [Mimeograph]

U. S. Federal Energy Administration, Office of International Energy Affairs. "Second Update of a Summary of Trade in Crude Oil between the U. S. and OPEC countries." Washington: Federal Energy Administration, 9 June 1975. [Mimeograph]

INSTITUTE OF INTERNATIONAL STUDIES
UNIVERSITY OF CALIFORNIA, BERKELEY

CARL G. ROSBERG,
Director

Monographs published by the Institute include:

RESEARCH SERIES

1. *The Chinese Anarchist Movement*, by Robert A. Scalapino and George T. Yu. ($1.00)
6. *Local Taxation in Tanganyika*, by Eugene C. Lee. ($1.00)
7. *Birth Rates in Latin America: New Estimates of Historical Trends*, by O. Andrew Collver. ($2.50)
12. *Land Tenure and Taxation in Nepal*, Volume IV, *Religious and Charitable Land Endowments: Guthi Tenure*, by Mahesh C. Regmi. ($2.75)
13. *The Pink Yo-Yo: Occupational Mobility in Belgrade, ca 1915-1965*, by Eugene A. Hammel. ($2.00)
14. *Community Development in Israel and the Netherlands: A Comparative Analysis*, by Ralph M. Kramer. ($2.50)
*15. *Central American Economic Integration: The Politics of Unequal Benefits*, by Stuart I. Fagan. ($2.00)
16. *The International Imperatives of Technology: Technological Development and the International Political System*, by Eugene B. Skolnikoff. ($2.95)
*17. *Autonomy or Dependence as Regional Integration Outcomes: Central America*, by Philippe C. Schmitter. ($1.75)
18. *Framework for a General Theory of Cognition and Choice*, by R.M. Axelrod. ($1.50)
19. *Entry of New Competitors in Yugoslav Market Socialism*, by S.R. Sacks. ($2.50)
*20. *Political Integration in French-Speaking Africa*, by Abdul A. Jalloh. ($3.50)
21. *The Desert and the Sown: Nomads in the Wider Society*, ed. by Cynthia Nelson. ($3.50)
22. *U.S.-Japanese Competition in International Markets: A Study of the Trade-Investment Cycle in Modern Capitalism*, by John E. Roemer. ($3.95)
23. *Political Disaffection Among British University Students: Concepts, Measurement, and Causes*, by Jack Citrin and David J. Elkins. ($2.00)
24. *Urban Inequality and Housing Policy in Tanzania: The Problem of Squatting*, by Richard E. Stren. ($2.50)
*25. *The Obsolescence of Regional Integration Theory*, by Ernst B. Haas. ($2.95)
26. *The Voluntary Service Agency in Israel*, by Ralph M. Kramer. ($2.00)
27. *The SOCSIM Demographic-Sociological Microsimulation Program: Operating Manual*, by Eugene A. Hammel et al. ($4.50)
28. *Authoritarian Politics in Communist Europe: Uniformity & Diversity in One-Party States*, ed. by Andrew C. Janos. ($3.95)
29. *The Anglo-Icelandic Cod War of 1972-1973: A Case Study of a Fishery Dispute*, by Jeffrey A. Hart. ($2.00)
30. *Plural Societies and New States: A Conceptual Analysis*, by Robert Jackson. ($2.00)
31. *The Politics of Crude Oil Pricing in the Middle East, 1970-1975: A Study in International Bargaining*, by Richard Chadbourn Weisberg. ($3.95)
32. *Agricultural Policy and Performance in Zambia: History, Prospects, and Proposals for Change*, by Doris Jansen Dodge. ($4.95)
33. *Five Classy Programs: Computer Procedures for the Classification of Households*, by E.A. Hammel and R.Z. Deuel. ($3.75)
34. *Housing the Urban Poor in Africa: Policy, Politics, and Bureaucracy in Mombasa*, by Richard E. Stren. ($5.95)
35. *The Russian New Right: Right-Wing Ideologies in the Contemporary USSR*, by Alexander Yanov. ($4.50)

*International Integration Series

INSTITUTE OF INTERNATIONAL STUDIES MONOGRAPHS (continued)

36. *Social Change in Romania, 1860-1940: A Debate on Development in a European Nation*, ed. by Kenneth Jowitt. ($4.50)
37. *The Leninist Response to National Dependency*, by Kenneth Jowitt. ($2.50)
38. *Socialism in Sub-Saharan Africa: A New Assessment*, ed. by Carl G. Rosberg and Thomas M. Callaghy. ($8.50)
39. *Tanzania's Ujamaa Villages: The Implementation of a Rural Development Strategy*, by Dean E. McHenry, Jr. ($5.95)
40. *Who Gains from Deep Ocean Mining? Simulating the Impact of Regimes for Regulating Nodule Exploitation*, by I.G. Bulkley. ($3.50)
41. *Industrialization, Industrialists, and the Nation-State in Peru: A Comparative/Sociological Analysis*, by Frits Wils. ($5.95)
42. *Ideology, Public Opinion, and Welfare Policy: Attitudes toward Taxes and Spending in Industrialized Societies*, by Richard M. Coughlin. ($4.95)
43. *The Apartheid Regime: Political Power and Racial Domination*, ed. by Robert M. Price and Carl G. Rosberg. ($8.95)

POLITICS OF MODERNIZATION SERIES

1. *Spanish Bureaucratic-Patrimonialism in America*, by Magali Sarfatti. ($2.00)
2. *Civil-Military Relations in Argentina, Chile, and Peru*, by Liisa North. ($2.00)
3. *Notes on the Process of Industrialization in Argentina, Chile, and Peru*, by Alcira Leiserson. ($1.75)
6. *Modernization and Coercion*, by Mario Barrera. ($1.50)
7. *Latin America: The Hegemonic Crisis and the Military Coup*, by José Nun. ($2.00)
8. *Developmental Processes in Chilean Local Government*, by Peter S. Cleaves. ($1.50)
9. *Modernization and Bureaucratic-Authoritarianism: Studies in South American Politics*, by Guillermo A. O'Donnell. ($5.50)

POLICY PAPERS IN INTERNATIONAL AFFAIRS

1. *Images of Detente and the Soviet Political Order*, by Kenneth Jowitt. ($1.00)
2. *Detente After Brezhnev: The Domestic Roots of Soviet Foreign Policy*, by Alexander Yanov. ($3.00)
3. *The Mature Neighbor Policy: A New United States Economic Policy for Latin America*, by Albert Fishlow. ($2.00)
4. *Five Images of the Soviet Future: A Critical Review and Synthesis*, by George W. Breslauer. ($2.50)
5. *Global Evangelism Rides Again: How to Protect Human Rights Without Really Trying*, by Ernst B. Haas. ($2.00)
6. *Israel and Jordan: The Implications of an Adversarial Partnership*, by Ian Lustick. ($2.00)
7. *Political Syncretism in Italy: Historical Coalition Strategies and the Present Crisis*, by Giuseppe Di Palma. ($2.00)
8. *U.S. Foreign Policy in Sub-Saharan Africa: National Interest and Global Strategy*, by Robert M. Price. ($2.25)
9. *East-West Technology Transfer in Perspective*, by R.J. Carrick. ($2.75)
10. *NATO's Unremarked Demise*, by Earl C. Ravenal. ($2.00)
11. *Toward an Africanized U.S. Policy for Southern Africa: A Strategy for Increasing Political Leverage*, by Ronald T. Libby. ($3.95)
12. *The Taiwan Relations Act and the Defense of the Republic of China*, by Edwin K. Snyder et al. ($3.95)

Address correspondence to:
   Institute of International Studies
   215 Moses Hall
   University of California
   Berkeley, California 94720